The MATHEMATICS of EVERYDAY LIFE

ALSO BY ALFRED S. POSAMENTIER, ROBERT GERETSCHLÄGER, CHARLES LI, AND CHRISTIAN SPREITZER

The Joy of Mathematics

ALSO BY ALFRED S. POSAMENTIER AND ROBERT GERETSCHLÄGER

The Circle

ALSO BY ALFRED S. POSAMENTIER AND INGMAR LEHMANN

The Fabulous Fibonacci Numbers

Pi: A Biography of the World's Most Mysterious Number

Mathematical Curiosities

Magnificent Mistakes in Mathematics

The Secrets of Triangles

Mathematical Amazements and Surprises

The Glorious Golden Ratio

ALSO BY ALFRED S. POSAMENTIER AND BERND THALLER

Numbers

ALSO BY ALFRED S. POSAMENTIER

The Pythagorean Theorem

Math Charmers

The MATHEMATICS of EVERYDAY LIFE

Alfred S. Posamentier and Christian Spreitzer

59 John Glenn Drive
Amherst, New York 14228

Published 2018 by Prometheus Books

The Mathematics of Everyday Life. Copyright © 2018 by Alfred S. Posamentier and Christian Spreitzer. All rights reserved. No part of this publication may be reproduced, stored in a retrieval system, or transmitted in any form or by any means, digital, electronic, mechanical, photocopying, recording, or otherwise, or conveyed via the internet or a website without prior written permission of the publisher, except in the case of brief quotations embodied in critical articles and reviews.

Trademarked names appear throughout this book. Prometheus Books recognizes all registered trademarks, trademarks, and service marks mentioned in the text.

The Internet addresses listed in the text were accurate at the time of publication. The inclusion of a website does not indicate an endorsement by the author(s) or by Prometheus Books, and Prometheus Books does not guarantee the accuracy of the information presented at these sites.

Cover images © Shutterstock
Cover design by Nicole Sommer-Lecht
Cover design © Prometheus Books

Unless otherwise indicated, all interior images are by the authors.

Inquiries should be addressed to
Prometheus Books
59 John Glenn Drive
Amherst, New York 14228
VOICE: 716–691–0133
FAX: 716–691–0137
WWW.PROMETHEUSBOOKS.COM

22 21 20 19 18   5 4 3 2 1

Library of Congress Cataloging-in-Publication Data

Names: Posamentier, Alfred S., author. | Spreitzer, Christian, 1979- author.
Title: The mathematics of everyday life / by Alfred S. Posamentier and Christian
    Spreitzer.
Description: Amherst, New York : Prometheus Books, 2018. | Includes index.
Identifiers: LCCN 2017042583 (print) | LCCN 2017057078 (ebook) |
    ISBN 9781633883888 (ebook) | ISBN 9781633883871 (hardcover)
Subjects: LCSH: Mathematics—Popular works.
Classification: LCC QA93 (ebook) | LCC QA93 .P6725 2018 (print) | DDC 510—dc23
LC record available at https://lccn.loc.gov/2017042583

Printed in the United States of America

*To my children and grandchildren, whose future is unbounded,
Lisa, Daniel, David, Lauren, Max, Samuel, Jack, and Charles*
—*Alfred S. Posamentier*

*To my beloved parents and grandmothers,
who continue to offer guidance and motivation,
Brigitte and Maximillian Spreitzer, as well as
Hildegard Gamauf and Anna Spreitzer*
—Christian Spreitzer

CONTENTS

INTRODUCTION	11

CHAPTER 1: HISTORICAL HIGH POINTS IN THE DEVELOPMENT OF MATHEMATICAL APPLICATIONS — 15

The Origin of Our Number Symbols	15
The Most Important Number in Mathematics	17
The Famous Fibonacci Numbers	20
Arithmetic in Ancient Egypt	21
Where the Terms Related to Our Clock Evolve	28
A Minute History of Timekeeping	28
Babylonian Mathematics and the Sexagesimal System	29
Babylonian Minutes and Seconds Have Survived to This Day	33
Roman Numerals Are Everywhere around Us	34
Mathematics on the Calendar	39
How We Overlook Our Calendar	48

CHAPTER 2: MATHEMATICS IN OUR EVERYDAY LIVES—ARITHMETIC SHORTCUTS AND THINKING MATHEMATICALLY — 53

Arithmetic with the Numbers 9 and 11	54
How 9s Can Check Your Arithmetic	60
Rules for Divisibility	62
A Quick Method to Multiply by Factors of Powers of 10	68

Arithmetic with Numbers of Terminal Digit 5	69
Multiplying Two-Digit Numbers Less Than 20	71
Mental Arithmetic Can Be More Challenging—but Useful!	73
Arithmetic with Logical Thinking	74
Using the Fibonacci Numbers to Convert Kilometers to and from Miles	75
Thinking "Outside the Box"	79
Solving Problems by Considering Extremes	82
The Working-Backward Strategy in Problem Solving	86

CHAPTER 3: MATHEMATICAL APPEARANCES AND APPLICATIONS IN EVERYDAY-LIFE PROBLEMS 93

Shopping with Mathematical Support	93
Successive Percentages	94
Raising Interest!	100
The Rule of 72	104
Paper Sizes and the Root of All ISO	106
Comparing Areas and Perimeters	110
Mathematics in Home Construction	113
The Perfect Manhole Cover	117
Design Your Own Coffee-Cup Sleeve!	124
How to Optimally Wrap a Present	128

CHAPTER 4: PROBABILITY, GAMES, AND GAMBLING 135

Friday the Thirteenth!	135
Unexpected Birthday Matches	137
Selecting Clothes	141
Playing Cards, a Counterintuitive Probability	142
Mathematics in Poker	144

Mathematical Logic of Tic-Tac-Toe 150
The Monty Hall Problem 154
Business Applications 158
Mathematics of Life Insurance 163
The Most Misunderstood Average 168
What We Need to Know about Averages 170
Comparing Measures of Central Tendency 172

CHAPTER 5: SPORTS AND GAMES— EXPLAINED MATHEMATICALLY 181

The Best Angle to Throw a Ball 181
Optimizing Your Shot at Soccer 187
A Game of Angles 192
Playing Billiards Cleverly 201
Mathematics on a Bicycle 206
The Spirograph Toy 212

CHAPTER 6: THE WORLD AND ITS NATURE 225

Measures of and on the Earth 225
Navigating the Globe 229
What Is Relativity? 232
Coloring a Map 233
Crossing Bridges 237
Mathematics in Nature 243
The Male Bee's Family Tree 244
Fibonacci Numbers in the Plant World 246
The Pine Cone and Others 247
Leaf Arrangement—Phyllotaxis 251
The Fibonacci Numbers on the Human Body 255
The Geometry of Rainbows 258

CHAPTER 7: APPEARANCES OF MATHEMATICS IN ART AND ARCHITECTURE — 275

- Golden Ratio Sightings — 276
- Displaying a Watch — 281
- Applications in Art — 284
- Perspectivity in Art — 296
- Numbers in Art — 305
- Viewing a Statue Optimally — 309
- The Most Overlooked Curve — 312
- The One-Sided Belt—the Möbius Strip — 315

CHAPTER 8: THE TECHNOLOGY AROUND US— FROM A MATHEMATICAL PERSPECTIVE — 319

- A Fascination with the Clock — 319
- The Mathematics of Paper Folding — 322
- Building a Skewed Tower — 328
- Whispering Galleries — 334
- Looking inside a Flashlight — 345
- Coffee with Caustics — 351
- Green Traffic Lights All the Way — 356
- Safety in Numbers — 365
- The ISBN System — 376
- How the Global Positioning System (GPS) Works — 381

ACKNOWLEDGMENTS — 391

NOTES — 393

INDEX — 401

INTRODUCTION

We hear so often that mathematics is important for us to better appreciate the world around us. This is the book that will guide you through a wide variety of aspects of our world that are accompanied by mathematical explanations. Unfortunately, because of restricted curriculum guidelines, your teachers and school probably did not have the time to navigate the many excursions that we will be presenting here. Had they been able to enrich mathematics instruction through these many applications, your math instruction would have been much more enjoyable. Teachers are strictly directed by curriculum guidelines to cover topics considered essential foundation blocks for further study in mathematics as well as the other STEM fields. This is compounded even more so in many schools where teachers are rated by the scores that their students achieve on standardized tests. Hence, there is a lot of "teaching to the test" and very little beyond that. In this book, we plan to dispel the very popular notion that mathematics is tedious and boring. We will be presenting a host of topics and ideas you are unlikely to have encountered during your school instruction, and they will give you a true feeling of how mathematics is a part of our everyday lives and how it can be used to explain many of the concepts we experience, the things we see, the decisions we make, and the overall understanding we have of the world around us.

Most people experience mathematics in a formal sense during their school years. All too often, mathematics is presented as a collection of mechanical techniques that supposedly allow the learner to apply these methods and concepts to a variety of encounters in everyday life. But the lifestyles and interests of the general populace vary broadly. Compound that with the lack of applications in the curriculum, and the average person is left with the feeling that he or she learns mathematics because it had to be learned and not because it is useful. In this book, we hope to convince you not only that mathematics is useful, but also that it can help us explain many things in everyday life that we tend to take for granted—and in unexpected ways.

There are people who feel that they can explain the world through mathematics. Oftentimes there may be a bit of a "stretch" in such explanations. Yet there are numerous parts of our experiential world that are based on mathematical concepts and principles. These can be appreciated in a variety of forms of art, architecture, nature, and finance; further, the daily aspects of our culture and lifestyles have mathematical explanations, or even mathematical origins. Most people are unaware as they confront these things that simple mathematics is being applied or can be used to better explain what is being observed.

As mentioned above, mathematical applications can be found in art and architecture, where the proportions of pictures, their subjects, and the nature of the structures themselves can be explained with well-established mathematical relationships. The golden ratio, which shows its tested beauty on the dimensions of the golden rectangle, is used throughout art and architecture. The structures that have for centuries become icons in our society—such as famous cathedrals, the Parthenon, the pyramids in Egypt—and other notable structures, even those built in modern times—such as the Headquarters of the United Nations in New York—all exhibit the beautiful golden ratio.

Additionally, for paintings to convey proper depth perception, lines of perspectivity also must come into play; we can see this in Leonardo da Vinci's famous *The Last Supper* mural in Milan, Italy, for example. Through their drawings and paintings, well-known artists have demonstrated their awareness of mathematical concepts. Notably among these is the famous German artist Albrecht Dürer, who studied Leonardo's techniques and used them in some of his drawings. He also showed a keen awareness of mathematics in his 1514 etching *Melencolia I*, in which he included an incredibly rich magic square that has many properties beyond those of normal magic squares. We hope to give you the tools you need to appreciate art and architecture from a mathematical standpoint, as well as an aesthetic one.

Perhaps the most ubiquitous numbers in mathematics are the Fibonacci numbers, which continuously arise in just about every aspect of our lives. For example, in nature the numbers of spirals on pine cones and pineapples are always Fibonacci numbers, as is the arrangement of the branches on a pear tree. You'll have to read on to learn about even more amazing appearances and applications of the Fibonacci numbers. (By the way, the Fibonacci numbers can also generate the golden ratio!)

Beyond art and the natural world, there are also unexpected, curious mathematical explanations for understanding excellence in sports. For example, you can use very simple high-school geometry to determine the optimal point along the sideline of a soccer field from which to shoot a goal. You can also use simple geometry in billiards to easily determine the point at which to hit a ball with the cue so that it hits a cushion and ricochets onto another cushion or ball. You might not think about this when playing a game of billiards for fun with your friends, but this is just another example of how math is all around you.

As you know, we all do lots of calculating and estimating in the course of our everyday lives. But you will be surprised, and perhaps entertained, by the many shortcuts and unusual relationships that can be used to make these tasks almost trivial, such as when converting miles to and from kilometers. We will present a variety of useful shopping shortcuts, investment insights, even how best to wrap a present!

Mathematics can help you navigate the globe and even appreciate and understand rainbows and the other curves that we encounter. For instance, when you travel along a road with timed traffic lights, mathematics can explain how this is done. Have you ever wondered why all sewer covers are round? That, too, will be explained. There are curves that enable us to have whispering galleries, and curves that hold up bridges. All of these are special properties that can be explained very easily through elementary mathematics.

The field of probability allows us insight into some unusual aspects of reality, too. It is clearly to a gambler's advantage to understand concepts of probability, for oftentimes a correct assessment of a wagering situation can be quite counterintuitive. There are game shows on television, most notably *Let's Make a Deal*, that have been a hot topic of controversy regarding how to determine the best strategy to win the game. Probability can also affect your worldview, particularly when you are reading a newspaper and journalists enthusiastically offer statistical evidence to support a position; at this point, knowledge of probability concepts can be helpful not only to understand the presented material but also to criticize it intelligently. We hope to enlighten you in this regard. We will also provide some curious insights into the card game of poker.

As you will see, mathematical problem-solving strategies are often used in everyday life. For example, using extremes to solve certain problems can be very effective not only in mathematics, but also with issues we face regularly. When confronted with a decision to be made, we say to ourselves, "Well, in the worst-case scenario, such and such would be the case." This allows us to move ahead with a sensible procedure for dealing with the situation at hand. Mathematical problem-solving strategies can guide us in the way we think about common, everyday decisions to be made.

These are just a few of the plethora of mathematical applications in our everyday lives. Often, we are not even aware that mathematics can explain and facilitate an understanding of what we see and how we can best deal with these situations. The mathematical concepts we consider in this book will require nothing more than a recollection of what you had been taught up to the tenth grade in high school. Join us now as we begin our journey through an investigation of a wealth of topics that either depend on mathematics or can use mathematics to explain their functioning, or, perhaps, even allow us to appreciate the world around us in an enhanced, sophisticated fashion because now many things we may have just accepted without question will become more meaningful.

CHAPTER 1
HISTORICAL HIGH POINTS IN THE DEVELOPMENT OF MATHEMATICAL APPLICATIONS

As we begin our exploration of the many obvious, and less obvious, occurrences of mathematics in our daily experiences, it would only be fitting that we begin with some of the historical highlights of where our mathematical usage began. It is well-known that the earliest uses of mathematics were portrayed with symbols of different kinds than what is used today. Best known is the use of hieroglyphics in the ancient Egyptian world. We still see today many applications of roman numerals—on cornerstones of buildings, on some clock or wristwatch faces, some preliminary book pages, and so on. Yet the most common use of symbols is the set of the 10 digits we use to represent numbers in the base-10, which originated in India. Let's first consider the origin of numbers so that we can better appreciate those that we use today.

THE ORIGIN OF OUR NUMBER SYMBOLS

Concepts for the numbers 1, 2, 3, 4, 5, 6, 7, . . . can be traced back to the origins of human history. Representations of these numbers must have arisen naturally as soon as humans started to count objects, animals,

16 THE MATHEMATICS OF EVERYDAY LIFE

or members of a clan. While finger-counting might have been the first counting method in history, recording the result requires the use of a permanent representation of the corresponding number, such as putting down a mark for each object counted. Representing a certain number by a corresponding number of marks (for example, notches on a wooden stick or on a bone) will eventually lead to the invention of numerals. Tally marks are the most primitive and oldest numeral system (see fig. 1.1). They are still used today for counting or tallying the score in a game.

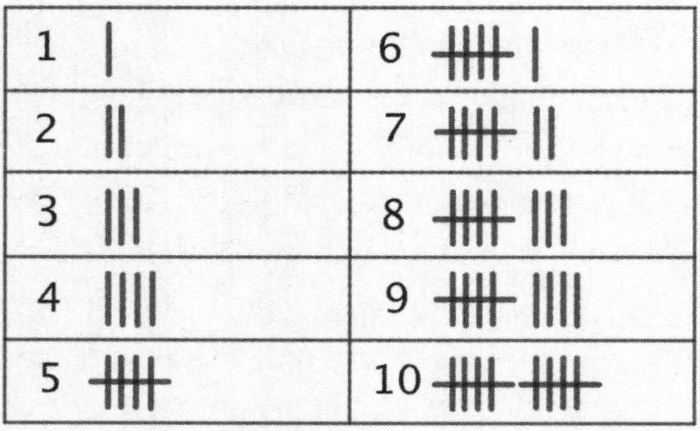

1	|	6	⊞ |
2	||	7	⊞ ||
3	|||	8	⊞ |||
4	||||	9	⊞ ||||
5	⊞	10	⊞ ⊞

Figure 1.1.

Since the representation of any number is obtained from that of its predecessor by simply adding one mark, tally marks are very convenient for documenting ongoing results. Historically, they constitute the origin of all numeral systems, and their usage dates back to the Upper Paleolithic age (roughly 50,000 to 10,000 BCE). References to tally marks can also be found in much more sophisticated numeral systems of ancient cultures, and the Japanese symbols for 1, 2, and 3 bear this resemblance still today (see fig. 1.2).

HISTORICAL HIGH POINTS

Figure 1.2. (From left to right) Ancient Egyptian; Babylonian; Ancient Indian; Roman; and Japanese numeral systems.

The introduction of numerals for the quantities represented by 1, 2, 3, 4, 5, 6, 7, ... is an almost inevitable consequence of the act of counting. We then could say that the positive integers were not "developed" or "invented"; rather, they were "discovered." The German mathematician Leopold Kronecker (1823–1891) is often quoted with the phrase, "God made the natural numbers; all else is the work of man." Here, the term "natural number,"[1] referring to an element of the set N = {1, 2, 3, ... }, was introduced in 1888 by the German mathematician Richard Dedekind (1831–1916).

THE MOST IMPORTANT NUMBER IN MATHEMATICS

The first number that was truly "invented" by the human mind is the number zero, 0. Many mathematicians would regard zero as the most important number in mathematics, since a number system without a notation for zero is extremely impractical for dealing with very small or very large numbers. Just imagine how to represent one million or one thousandth as a decimal number without having a symbol for the concept of zero.

The development of a place-value system by the Babylonians in ancient Mesopotamia in the second millennium BCE was one of the first historical highpoints of mathematics. In a place-value system, the exact position of a digit within a number is relevant for the value it represents. The Babylonians had no concept or notation for zero, which caused difficulties for many-digit numbers and resulted in severe notational problems. Since their numeral system was a sexagesimal system (base-60), a Babylonian three-digit number $D_2 D_1 D_0$ would represent the number $D_2 \cdot 60^2 + D_1 \cdot 60^1 + D_0 \cdot 60^0$ in the decimal system, but the notational problems that arise when there is no zero digit are essentially the same for all place-value systems. To demonstrate these notational problems, we will consider familiar decimal numbers rather than the Babylonian sexagesimal numbers. If there is no digit 0, how can we, for example, distinguish the number 2018 from 218? The Babylonians indicated the absence of a certain positional value by a space between sexagesimal numerals. This would correspond to writing "2 18" for the number 2018. Clearly, this practice becomes quite problematic when two or more consecutive positions need to be left out, as in the number 1001. Moreover, using spaces for empty positions does not resolve all ambiguities, since we can hardly add a space at the end of a number. Indeed, the Babylonians used identical representations for 1, 60, and 60^2, which would correspond to writing just "1" for the numbers 1, 10, 100, or any other power of 10. The number that was actually intended always had to be inferred from the context. Around 300 BCE, the Babylonians introduced a placeholder symbol for empty positions, making the notation less ambiguous. Yet this symbol was never used at the end of a number, so the written numbers 1, 60, and 60^2 still could not be distinguished from one another without knowing the context.

Babylonian clay tablets dating from the second millennium BCE already cover calculations with fractions, algebra, quadratic and cubic equations, and the Pythagorean theorem—more than a thousand years before Pythagoras was born. It might appear strange to us that one of the most mathematically advanced civilizations of their time never developed a concept of zero, especially when we think about the associated notational difficulties the Babylonians had to deal with.

Who invented the zero, then? It is documented that the Greco-Egyptian mathematician and philosopher Claudius Ptolemy (ca. 100–ca. 170) used a symbol for zero within a sexagesimal numeral system in his astronomical writings.[2] However, the symbol usually played the role of a placeholder, and it was only used in the fractional part of a number, that is, the part corresponding to minutes and seconds. The modern concept of zero as an integral number was actually developed in India, together with the decimal place-value system. The oldest clear evidence for a thorough understanding of the role of zero was found in the text *Brāhmasphuṭasiddhānta*, which was the main work of the Indian mathematician and astronomer Brahmagupta (ca. 598–ca. 665).[3] Some older writings indicate that the number zero was used in India as early as the fifth century CE. In the ninth century, the concept of zero was transmitted to the Islamic culture and later brought to Europe by the famous Italian mathematician Fibonacci (ca. 1170–1250), who in his book *Liber Abaci* (1202) used the term *zephyrum* as a translation of the Arabic *ṣifr*, meaning "empty." The word *zephyrum* turned into *zefiro* in Italian and was then contracted to *zero* in Venetian.[4] (More about Fibonacci in the next section.)

The introduction of zero into the decimal system dramatically simplified calculations with large numbers, thereby making mathematics, in general, a much more practicable and powerful tool. It surely was the most significant step in the development of a number system, and it laid the basis for many great advances in commerce, navigation, astronomy, physics, and engineering. Inventing a notion of zero, and raising this symbol for "nothing" to a number on its own right, requires a considerable mental abstraction that should be ranked among the greatest achievements of the human mind in the ancient world.

THE FAMOUS FIBONACCI NUMBERS

Now that we have established the beginnings of the number zero, let's see how the numbers that we use today evolved and spread throughout the world. As mentioned above, they originated in India—invented there by mathematicians between the first and fourth centuries CE—and were brought into the European world in the year 1202 by the famous mathematician Leonardo of Pisa, better known today as Fibonacci. As mentioned above, Fibonacci wrote a book called *Liber Abaci*. It contained many mathematical problems, and the first words of its introduction were:

> The nine Indian figures are: 9 8 7 6 5 4 3 2 1.

> With these nine figures, and with the sign 0, which the Arabs call zephyr, any number whatsoever is written, as demonstrated below. A number is a sum of units, and through the addition of them the number increases by steps without end. First one composes those numbers, which are from one to ten. Second, from the tens are made those numbers, which are from ten up to one hundred. Third, from the hundreds are made those numbers, which are from one hundred up to one thousand. . . . And thus, by an unending sequence of steps, any number whatsoever is constructed by joining the preceding numbers. The first place in the writing of the numbers is at the right. The second follows the first to the left. . . .[5]

Fibonacci encountered these numbers during his travels on the Barbary Coast of Africa, where he worked closely with Arab mathematicians. Today, these numerals are referred to as Hindu-Arabic numerals, depicting their path to Europe. Despite their relative facility, these numerals were not widely accepted by merchants who were suspicious of those who knew how to use them. Such merchants were simply afraid of being cheated. The numerals first began to be used fifty years after the publication of Fibonacci's book, but still not very extensively. We can safely say that it took the same three hundred years for these numerals to catch on as it did for the leaning tower of Pisa to be completed.

Interestingly, *Liber Abaci* also contains simultaneous linear equations. Many of the problems that Fibonacci considers, however, were similar to those appearing in Arab sources. This does not detract from the value of the book, since it is the collection of solutions to these problems that makes it a major contribution to our development of mathematics. As a matter of fact, a number of mathematical terms—common in today's usage—were first introduced in *Liber Abaci*. Fibonacci referred to *factus ex multiplicatione*, and from this first sighting of the phrase, we now speak of the "factors of a number" or the "factors of a multiplication." Another example of words whose introduction into the current mathematics vocabulary seems to stem from this famous book are "numerator" and "denominator."

Before we go further with these common numerals, it is worthwhile to see what predated them. This we will see when we look at the ancient Egyptian number system.

ARITHMETIC IN ANCIENT EGYPT

In 1858, the young Scottish lawyer Alexander Henry Rhind (1833–1863) purchased a papyrus that was found during illegal excavations in the Theban Necropolis, which lies within the modern Egyptian city of Luxor. This papyrus turned out to be one of the oldest documents of written mathematics, and, to this day, it is the main source of our knowledge about ancient Egyptian mathematics. The papyrus dates to around 1650 BCE, and in the opening paragraph, the scribe states that he is copying a text from the reign of King Amenemhat III. The original text, which is now lost, must have been even a few hundred years older.

Figure 1.3.

Today, the so-called Rhind Mathematical Papyrus (fig. 1.3) is kept at the British Museum in London, but some small fragments are held by the Brooklyn Museum in New York. The text contains a collection of eighty-seven problems in arithmetic, algebra, and geometry. As a matter of fact, it showed that the ancient Egyptians had calculated the value of π as 3.1605, which is less than 1 percent off from the actual value.[6] The Rhind Papyrus gives us insight into the mathematical knowledge in ancient Egypt, as well as providing us with explicit calculations, such as demonstrating how multiplication and division were carried out. This is particularly interesting because of the way that computation methods in ancient civilizations differ from the modern algorithms for multiplication and division that we use today. The ancient Egyptians used a systematic method for multiplying two numbers that only requires the ability to multiply and divide by 2, and to add.

For instance, to find the product of 19 and 34 with this method, we would first create a two-column table, writing 1 in the first column and one of the multiplicands, say, 34, in the first row of the second column. (A multiplicand is a number that is going to be multiplied by another; in this example, then, the multiplicands are 19 and 34.) We obtain each line in the first column by doubling the preceding line. Thus, the first column will contain a sequence of powers of 2 and the second column will contain a doubling sequence of 34. We proceed until the sequence in the first column arrives at a number that allows us to get a sum from the numbers in the first column that is equal to the other multiplicand, 19. This would be the largest number in the representation of 19 into the sum of powers of 2, that is, $19 = 2^4 + 2^1 + 2^0 = 16 + 2 + 1$. We now highlight (in boldface) the corresponding rows in the table, as shown in table 1.1.

Powers of 2 Contained in 19	Doublings of 34
1	**34**
2	**68**
~~4~~	~~136~~
~~8~~	~~272~~
16	**544**
Sum: **19**	Sum: **646**

Table 1.1. Ancient Egyptian multiplication method.

By adding the appropriate terms (in boldface) in the doubling sequence, we find the full product, that is, $19 \cdot 34 = 544 + 68 + 34 = 646$. It does not matter which of the multiplicands we decompose into powers of 2. However, it might be wise to take the one with the shorter decomposition, since this implies that we have fewer numbers to add. Table 1.2 shows the alternative variant of computing the product 19 · 34. Here we list the powers of 2 in the first column until we reach a number that allows us to create a sum equal 34. Here we see that 32 + 2 = 34. Taking the corresponding numbers in the second column, we find that 38 + 608 = 46.

Powers of 2 Contained in 34	Doublings of 19
~~1~~	~~19~~
2	38
~~4~~	~~76~~
~~8~~	~~152~~
~~16~~	~~304~~
32	608
Sum: 34	Sum: 646

Table 1.2. Ancient Egyptian multiplication method.

A shorter decomposition of one multiplicand means that we have to double the other multiplicand more often (since a shorter decomposition will necessarily involve higher powers of 2), so there is actually not that much work to be saved.

The most remarkable fact about the ancient Egyptians' arithmetic is their very special way of calculating with fractions. Only fractions of the form $\frac{1}{n}$ were used, with the exception of $\frac{2}{3}$ (and, very rarely, $\frac{3}{4}$). There was not even a notation for other fractions! All fractions they had to deal with (typically as the results of divisions) were expressed as sums of unit fractions (each having numerator 1). Moreover, all of the unit fractions in such a sum had to be different! For example, the fraction $\frac{2}{5}$ was not represented as $\frac{1}{5}+\frac{1}{5}$, but as $\frac{1}{3}+\frac{1}{15}$. To see that this is equal to $\frac{2}{5}$, we rewrite $\frac{1}{3}+\frac{1}{15}$ with common denominators and compute the sum: $\frac{1}{3}+\frac{1}{15}=\frac{5}{15}+\frac{1}{15}=\frac{6}{15}=\frac{2}{5}$. The Rhind Papyrus contains a table of representations of $\frac{2}{n}$ as sums of distinct unit fractions for odd n between 5 and 101. Table 1.3 shows the first twenty-four unit-fraction expansions in the table on the Rhind Papyrus.

HISTORICAL HIGH POINTS

$\frac{2}{3}=\frac{1}{2}+\frac{1}{6}$	$\frac{2}{15}=\frac{1}{10}+\frac{1}{30}$	$\frac{2}{27}=\frac{1}{18}+\frac{1}{54}$	$\frac{2}{39}=\frac{1}{26}+\frac{1}{78}$
$\frac{2}{5}=\frac{1}{3}+\frac{1}{15}$	$\frac{2}{17}=\frac{1}{12}+\frac{1}{51}+\frac{1}{68}$	$\frac{2}{29}=\frac{1}{24}+\frac{1}{58}+\frac{1}{174}+\frac{1}{232}$	$\frac{2}{41}=\frac{1}{24}+\frac{1}{246}+\frac{1}{328}$
$\frac{2}{7}=\frac{1}{4}+\frac{1}{28}$	$\frac{2}{19}=\frac{1}{12}+\frac{1}{76}+\frac{1}{114}$	$\frac{2}{31}=\frac{1}{20}+\frac{1}{124}+\frac{1}{155}$	$\frac{2}{43}=\frac{1}{42}+\frac{1}{86}+\frac{1}{129}+\frac{1}{301}$
$\frac{2}{9}=\frac{1}{6}+\frac{1}{18}$	$\frac{2}{21}=\frac{1}{14}+\frac{1}{42}$	$\frac{2}{33}=\frac{1}{22}+\frac{1}{66}$	$\frac{2}{45}=\frac{1}{30}+\frac{1}{90}$
$\frac{2}{11}=\frac{1}{6}+\frac{1}{66}$	$\frac{2}{23}=\frac{1}{12}+\frac{1}{276}$	$\frac{2}{35}=\frac{1}{30}+\frac{1}{42}$	$\frac{2}{47}=\frac{1}{30}+\frac{1}{141}+\frac{1}{470}$
$\frac{2}{13}=\frac{1}{8}+\frac{1}{52}+\frac{1}{104}$	$\frac{2}{25}=\frac{1}{15}+\frac{1}{75}$	$\frac{2}{37}=\frac{1}{24}+\frac{1}{111}+\frac{1}{296}$	$\frac{2}{49}=\frac{1}{28}+\frac{1}{196}$

Table 1.3. Unit fraction expansions found in the Rhind Mathematical Papyrus.

The decompositions are not unique; for example, here are several such representations for the fraction $\frac{2}{9}$ as follows: $\frac{2}{9}=\frac{1}{6}+\frac{1}{18}=\frac{1}{5}+\frac{1}{45}=\frac{1}{8}+\frac{1}{12}+\frac{1}{72}$. It is known neither why the ancient Egyptians almost exclusively used unit fractions, nor how they found their representations. A partial answer to the first question could be that decompositions into unit fractions are sometimes very practical if a certain quantity of goods would have to be split into equal parts. For instance, one of the first problems on the Rhind Papyrus asks how to divide 7 loaves of bread between 10 men. Converting $\frac{7}{10}$ into a decimal number would not be of any use here. Cutting each loaf into 10 equal slices and giving 7 slices to each person produces a large number of slices and a lot of unnecessary crumbs as waste. However, by expanding $\frac{7}{10}$ into a sum of unit fractions, we obtain $\frac{7}{10}=\frac{1}{2}+\frac{1}{5}$. Hence, everyone gets half a loaf and an extra fifth of a loaf. This is a much more efficient way of sharing 7 loaves between 10 people!

Today, mathematicians use the term "Egyptian fraction" for any sum of several distinct unit fractions. Although we do not know how the ancient Egyptians found their unit-fraction representations, there exist various algorithms to expand an arbitrary fraction in terms of an

Egyptian fraction. The first published methods to construct Egyptian fractions were found in Fibonacci's book *Liber Abaci*. One of the methods he describes is now known as the "greedy algorithm for Egyptian fractions"[7] and we will demonstrate how it works by showing an example. Consider the fraction $\frac{5}{17}$. To get the first term in the expansion, we seek the largest unit fraction smaller than $\frac{5}{17}$. Therefore, we increase the denominator until the fraction becomes reducible to a unit fraction. The first denominator increases $\frac{5}{18}$ and $\frac{5}{19}$ are not reducible to unit fractions, but the next one, $\frac{5}{20} = \frac{1}{4}$. This gives us $\frac{1}{4}$ as the first term in our expansion. To get the next term, we subtract $\frac{1}{4}$ from $\frac{5}{17}$ and repeat the procedure; that is, we determine the largest unit fraction smaller than or equal to the remainder, $\frac{5}{17} - \frac{1}{4}$, then subtract and repeat "the greedy algorithm," so called because one always takes away the largest unit fraction contained in the remainder. Here we would obtain $\frac{5}{17} - \frac{1}{4} = \frac{20}{68} - \frac{17}{68} = \frac{3}{68}$. Continuing our earlier procedure, we find the next higher-denominator fraction, $\frac{3}{69} = \frac{1}{23}$, which is the largest unit fraction we can subtract from the remainder. This leads to $\frac{3}{68} - \frac{1}{23} = \frac{69}{1564} - \frac{68}{1564} = \frac{1}{1564}$, and now we are done, since the remainder is already a unit fraction. We obtained the Egyptian fraction expansion $\frac{5}{17} = \frac{1}{4} + \frac{1}{23} + \frac{1}{1564}$. The algorithm will always come to an end, since each step reduces the numerator of the remaining fraction to be expanded. However, this method does not necessarily produce the shortest possible expansion. Actually, it will often produce expansions that are longer than the ancient Egyptians' expansions, or it will contain larger denominators. For example, using the greedy algorithm to expand $\frac{2}{21}$ yields $\frac{1}{11} + \frac{1}{231}$, involving a much larger denominator than the ancient Egyptian representation $\frac{1}{14} + \frac{1}{42}$. Interestingly, no general algorithm is known for producing unit-fraction representations that have either a minimum number of terms or the smallest possible denominators.

For some special cases there exist shortcuts to obtain a unit-fraction expansion. One of these special situations is when you are given an irreducible fraction with numerator 2. An irreducible fraction of the form $\frac{2}{n}$ can always be written as $\frac{2}{pq}$, where p and q are odd numbers, one of which may be 1 (if one or both of them were even, then $\frac{2}{pq}$ would be reducible). We then have the decomposition

$$\frac{2}{pq} = \frac{1}{\frac{p(p+q)}{2}} + \frac{1}{\frac{q(p+q)}{2}},$$

where $\frac{p+q}{2}$ must be an integer, since $p + q$ is even. One way to see this is to divide the identity $1 = \frac{p}{p+q} + \frac{q}{p+q}$ by pq, leading to $\frac{1}{pq} = \frac{1}{p(p+q)} + \frac{1}{q(p+q)}$, which yields the proposed decomposition upon multiplication by 2 (which we represent on the right-hand side of the equation as dividing the denominator by 2). Notice that since we may also multiply by another number than 2, we would get an analogous decomposition for fractions of the form $\frac{n}{pq}$, whenever $p + q$ is divisible by n. This method will always work for irreducible fractions with numerator 2. As an example, we may consider $\frac{2}{21} = \frac{2}{3 \cdot 7}$, for which the decomposition yields

$$\frac{2}{3 \cdot 7} = \frac{1}{3 \cdot \frac{10}{2}} + \frac{1}{7 \cdot \frac{10}{2}} = \frac{1}{15} + \frac{1}{35}.$$

You may have noticed that this is already the third Egyptian fraction expansion for $\frac{2}{21}$ presented above. Actually, there always exist infinitely many Egyptian fraction expansions for any given fraction. To see this, notice that every unit fraction can itself be expanded into other unit fractions by means of the identity

$$\frac{1}{n} = \frac{1}{n+1} + \frac{1}{n(n+1)}.$$

For example, $\frac{1}{2} = \frac{1}{3} + \frac{1}{6}$, and $\frac{1}{3} = \frac{1}{4} + \frac{1}{12}$, and so on. Thus, using this identity, we can construct arbitrarily long chains of unit fractions equal to a given fraction. This implies that there must exist infinitely many Egyptian fractions representing the same fractional number. A much more difficult question to ask is: How many Egyptian fractions of a given length does a given fractional number have? In 1948, the famous Hungarian mathematician Paul Erdös (1913–1996), together with the

German-American mathematician Ernst Straus (1922–1983),[8] conjectured that for every integer $n \geq 2$, the fraction $\frac{4}{n}$ has an Egyptian fraction expansion of length 3, meaning that there exist three positive integers a, b, and c such that $\frac{4}{n} = \frac{1}{a} + \frac{1}{b} + \frac{1}{c}$. This is one of several open problems on Egyptian fractions in number theory. It is quite charming that the ancient Egyptians' curious way of representing fractions generates some mathematical problems that are still unsolved today.

WHERE THE TERMS RELATED TO OUR CLOCK EVOLVE

Have you ever wondered why an hour is divided into 60 minutes and a minute into 60 seconds? The word "minute" comes from the Latin *pars minuta prima*, meaning "first small part," and the word "second" has its origin in the Latin *pars minuta secunda*, meaning "second small part." Thus, a minute is the "primary minute division" of an hour, and it is further divided into seconds, the "secondary minute divisions."

A MINUTE HISTORY OF TIMEKEEPING

In fact, minutes were not considered by the general public until the end of the seventeenth century. Although some clocks built in the fifteenth century already indicated minutes and seconds, such elaborate instruments were very rare, and their accuracy was poor. Most clocks from that time only had an hour hand. In 1656, the Dutch mathematician and physicist Christiaan Huygens (1629–1695) invented the pendulum clock, a crucial step in the history of timekeeping devices.[9] He was able to build clocks that erred by less than one minute per day. The accuracy of pendulum clocks was further improved by the invention of the anchor escapement, probably by the British physicist Robert Hooke (1635–1703).[10] An *escapement* is a mechanism that turns the clock's wheels by a fixed angle with each swing of the pendulum, thereby moving the clock's hands forward. The anchor escapement reduced the required amplitude of the swing of the pendulum from about 100° to only 5°, thus allowing for pendulums to be much longer and to swing at a slower rate. With the resulting improvement in accuracy, around 1680–1690 the minute hand became standard in pendulum clocks. Finally, the invention of the balance spring, attributed to both Christiaan Huygens and Robert Hooke, made it possible to build reasonably accurate and slim clocks, enabling the production of pocket watches displaying minutes, and eventually also seconds.[11]

But why is an hour divided into 60 minutes and not, as might be expected, 10 or 100 minutes, using numbers of our base-10 system? Although the historical origin of a division into 60 parts could never be fully clarified, it most probably derives from the sexagesimal system, a numeral system based on the number 60 that was developed in Mesopotamia (modern Iraq) around 2000 BCE. Before we explore the clock further, let's examine the sexagesimal system.

BABYLONIAN MATHEMATICS AND THE SEXAGESIMAL SYSTEM

Any mathematics that originated in the ancient Mesopotamia is now referred to as Babylonian mathematics, deriving its name from the city of Babylon, which was an intellectual center and important place of study from its rise around 1900 BCE to its fall in 539 BCE. The Babylonians wrote on clay tablets, which were baked in an oven or dried in the sun. Some of the earliest evidence of written mathematics is Plimpton 322 (see fig. 1.4), a Babylonian clay tablet in the Plimpton Collection at Columbia University in New York.

30 THE MATHEMATICS OF EVERYDAY LIFE

Figure 1.4.

The tablet was written in cuneiform script and lists what are now called Pythagorean triples, that is, integers a, b, and c satisfying the famous Pythagorean theorem: $a^2 + b^2 = c^2$. The Babylonian numeral system was a sexagesimal system (again, based on the number 60), and, as far as we know today, it was the first true place-value system in the history of mathematics. Today we are using a decimal system, a place-value system based on the number 10, not the number 60. If a place-value system is used to represent a number, the value of a particular digit depends on both the digit itself and its position within the number (for example, $225 = 2 \cdot 100 + 2 \cdot 10 + 5 \cdot 1$). Without a place-value system, each power of the base has to be represented by a unique symbol, making calculations quite cumbersome. The Romans, for instance, had numerals based on 10, but they did not use a place-value system and denoted 1, 10, 100, and 1000 by the symbols I, X, C, and M (as well as one-half of 10 with a V, one-half of 100 with an L, and one-half of 1,000 with a D).

In the Babylonian system, the same symbol was used for the numbers 1, 60, 60², 60³, and so on. The position of the symbol within a number determined which power of 60 it represented. This is an improvement when compared to the Roman system, for example, but it still had its drawbacks. For example, since the Babylonians did not have a concept of the number zero (and hence also no digit for zero), the very same symbol could stand for 1 or 60 or any other power of 60, depending on the context. Knowing the context of a mathematical text was, therefore, crucial for interpreting numbers correctly.

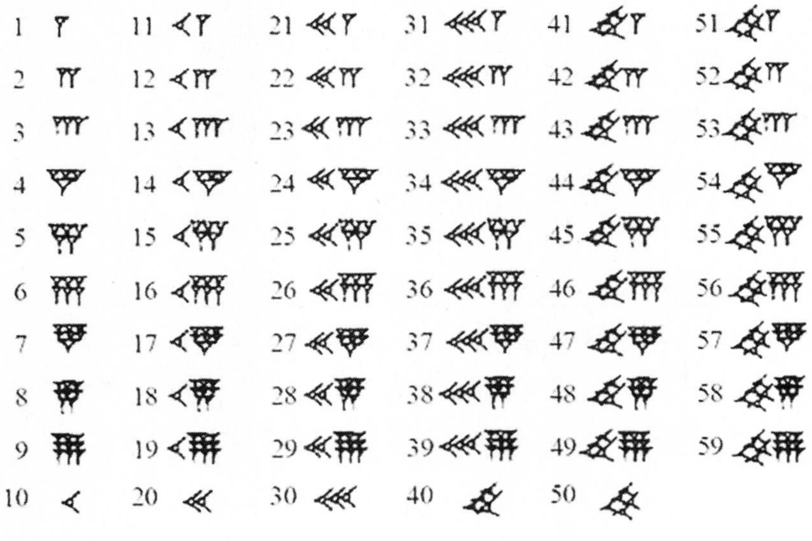

Figure 1.5.

As can be seen in figure 1.5, all 59 Babylonian non-zero digits were composed of at most two symbols. In order to facilitate the notation for the 59 different digits needed, an internal decimal system was used, similar to roman numerals (see fig. 1.5), but the place values in a digit string were always based on 60. The internal decimal notation for digits indicates that the Babylonians might also have experimented with a decimal place-value system before they decided to work with a sexagesimal system. But what is so special about the number 60? Why did the Babylonians base their arithmetic on 60 and not on 10 or any other number? In contrast to 10, the number 60 is a so-called highly composite number, which is a positive integer that has more divisors than any other smaller positive integer. The number 60 is divisible by 2, 3, 4, 5, 6, 10, 12, 15, 20, and 30. It is also the smallest number divisible by the first six counting numbers. This implies that the corresponding fractions $\frac{1}{2}, \frac{1}{3}, \frac{1}{4}, \frac{1}{5}, \frac{1}{6}$ can all be expressed by a single digit in the sexagesimal place-value system, whereas in the decimal system, only the fractions $\frac{1}{2}=0.5$ and $\frac{1}{5}=0.2$ have such a simple representation. The fractions $\frac{1}{3}=0.\overline{3}$ and $\frac{1}{6}=0.1\overline{6}$ do not even have a finite decimal representation. The divisibility of 60 by 1, 2, 3, 4, 5, and 6 was very helpful for trading, especially since ancient Mesopotamian units of measurement were based on integer factors of 60. Table 1.4 lists the basic units of length, area, and mass, together with their respective ratios.

Basic Length		Basic Area		Basic Mass	
finger	1/30	shekel	1/60	grain	1/180
foot	2/3	garden	1	shekel	1
cubit	1	quarter-field	5	pound	60
step	2	half-field	10	load	3600

Table 1.4.

Considering mass units, for instance, the fractional parts $\frac{1}{2}, \frac{1}{3}, \frac{1}{4}, \frac{1}{5}, \frac{1}{6}$ of a pound would correspond to 30, 20, 15, 12, and 10 shekels. Similarly, the fractional parts $\frac{1}{2}, \frac{1}{4}, \frac{1}{5}$, of a foot would correspond to 10, 5, and 4 fingers. It is quite obvious that the sexagesimal system is notably convenient for expressing fractions.

BABYLONIAN MINUTES AND SECONDS HAVE SURVIVED TO THIS DAY

Let's return to our inspection of the clock. The divisibility properties of 60 are also practical for measuring time, given that a day is divided into two 12-hour cycles (stemming from the ancient Egyptians). A 12-hour clock dial can be evenly divided into 60 minutes and can, therefore, be used to simultaneously display hours and minutes. Interestingly, during the French revolution in 1793, the French republicans wanted to introduce a decimal time system, dividing the day into 10 decimal hours, each consisting of 100 decimal minutes, which were divided into 100 decimal seconds.[12] However, the plan was abandoned a few years later, since it was not very well received by the general public. France started another attempt toward a decimalization of time measures in 1897, when the *Commission de décimalisation du temps* was founded, whose secretary was the famous mathematician Henri Poincaré (1854–1912).[13] This time a compromise was pursued, retaining the 24-hour day, but dividing each hour into 100 decimal minutes, and each minute into 100 seconds. Once again, this undertaking was given up a few years later, since it did not gain very much acceptance. As a consequence, to the present day, the sexagesimal system has survived as our measurement of time. Sticking to a system to which we are all accustomed might be just a force of habit, but perhaps the nice and unique divisibility properties of 60 somehow appeal to us as they must have to the Babylonians.

We are all aware that there is another numeral system that still plays a minor role in our society and was originally developed by the Romans. Since we encounter it in our everyday lives, we ought to take note of how it evolved and survived so long.

ROMAN NUMERALS ARE EVERYWHERE AROUND US

Football fans may have noticed that the 50th Super Bowl, played in February 2016, was branded with a logo that looked quite different from the Super Bowl logos that were used in prior years. While Super Bowls are usually numbered in roman numerals (a practice established at Super Bowl V in 1971), the National Football League broke with that tradition and went with arabic numerals for Super Bowl 50. In roman numerals, the letter L represents the number 50, so it would have been "Super Bowl L," using roman numerals. The NFL explained that the primary reason for the change was the difficulty in designing an aesthetically pleasing logo for "Super Bowl L."[14] But why did they use roman numerals in the first place, more than 1500 years after the fall of the Roman Empire?

Although roman numerals survived the Roman Empire by almost a millennium and remained common in Europe throughout the Middle Ages, they were eventually almost completely replaced by the more convenient Hindu-Arabic numerals during the Renaissance in Europe. However, we all know that in certain contexts, roman numerals are still being used today. Hour marks on clock faces are very often labeled with roman numerals; in the credits of films, the year of production is frequently written in roman numerals; and roman numerals can be seen on the corner stones of public buildings, monuments, and gravestones. Monarchs are usually numbered in roman, for example, Henry VIII and Elizabeth II of the United Kingdom. Roman numerals are also used as generational suffixes for persons who share the same name within a family, such as Harry Phillips III. You will find roman numerals on the front page of the *New York Times*, indicating the volume number; and as lower-case letters they are sometimes used for the preliminary pages of books before the main page numbering begins. The most prominent use of roman numerals in the world of sports is for the modern Olympic Games. For instance, the 1996 Summer Olympics in Atlanta, Georgia, are officially known as the "Games of the XXVI Olympiad." The numbering is useful both to distinguish between different events and also to know how many Olympic Games have already taken place (since the first modern Olympic Games in 1896). This was also the reason for numbering the Super Bowl events, and roman numerals were probably used to give them more of an "official character" and make the game seem more prestigious and dignified, comparable to the Olympic Games. However, unlike the Olympic Games, the Super Bowl is an annual event, and larger numbers tend to get longer and more complex to read when using roman numerals. So, in spite of the fact that the NFL returned to roman numerals for Super Bowl LI in 2017, this might not be the last word spoken on this issue.

Let's examine how roman numerals function. They are based on seven symbols, each of which represents a fixed value:

I	V	X	L	C	D	M
1	5	10	50	100	500	1000

Numbers are formed by combining these symbols and adding the values. It might be interesting to note that the symbol for 5 is V, which is the top half of X—the symbol for 10. Symbols are placed from left to right in order of value, with the largest at the beginning. To convert a number into Roman numerals, we remove the largest roman values, write down the corresponding numerals, and continue with the remainder until the entire value is converted. As an example, consider 476, the year in which the regency of the last Roman emperor ended. The largest roman values we can begin with are four Cs, so we write down CCCC, representing 400, which leaves a remainder of 76. Now we can cover the 50 from the remaining 76 with L, to obtain CCCCL, representing 450. The remainder to be covered is 26, which can be written as two Xs (for 20), and VI (for 5 plus 1, or 6); so, 26 is written as XXVI. Thus, we arrive at CCCCLXXVI, which is 476 written in roman numerals! Conversion to an ordinary number is easy; we just have to add up the values of the letters, going from left to right.

The Romans did not have a place-value system, where the value of a numeral would depend on its position within the number. Since the numerals had definitive values that were simply added up, larger numbers generally needed more symbols. To shorten the notation, subtractive rules were introduced:

- I before V or X indicates one fewer; for example, IV (one less than five: 4) or IX (one less than ten: 9)
- X before L or C indicates ten fewer; for example, XL (ten less than fifty: 40) or XC (ten less than one hundred: 90)
- C before D or M indicates one hundred fewer; for example, CD (one hundred less than five hundred: 400) or CM (one hundred less than one thousand: 900)

Thus, the year 476 can also be written as CDLXXVI, which is unambiguous since "CD" must be subtractive notation, since the order of the letters is lower to higher when read from left to right. Although subtractive notation is the usual method of writing Roman numbers today, the ancient Romans seem to have preferred additive forms such as IIII (four) or VIIII (nine), especially in written documents. Subtractive notation was employed only when there wasn't enough space or when the numerals had to be engraved in stone. In fact, subtractive notation became popular in the Middle Ages, several hundred years after the fall of the Roman Empire. Already in the Roman era both notations lived in peaceful coexistence, and, sometimes, additive and subtractive forms have been used in the very same document. This inconsistency has survived to this day, on clock faces. Many clock faces using Roman numerals show IIII for four o'clock, but IX for nine o'clock (subtractive notation), as shown in figure 1.6.

HISTORICAL HIGH POINTS   37

Figure 1.6.

Adding roman numbers is particularly simple if the numbers are written in additive form. For example, if we want to add 37 and 24—that is, XXXVII and XXIIII—we just to have to put all numerals together, group them, and convert to larger-value numerals if possible. Thus, the sum would be XXXXXVIIIIII, which, upon combining five Xs to an L, five Is to a V, and two Vs to an X, yields LXI, which is 61. Subtraction is also not difficult; we only have to cross out common numerals. However, we may need to convert larger-value numerals to smaller-value numerals in order to cross out all numerals of the subtracted number. For example, to subtract 24 from 61—that is, LXI minus XXIIII—we would convert LXI to XXXXXVIIIII and cross out the numerals forming the subtracted number (XXIIII), leaving us with XXXVII, which is indeed 37 (we basically apply the addition algorithm backward). The ancient Romans never developed a concept of zero or negative numbers, which might have something to do with this crossing-out procedure for subtraction. It is hard to make sense of crossing out something that is not there.

For very large numbers, the notation really got bulky. During the second century CE, the city of Rome already had more than one million inhabitants, making an additive number system consisting of only seven letters not a very convenient choice for a population census (one million would correspond to a sequence of thousand Ms).[15] Therefore, the Romans developed several special notations to express very large numbers. We will just mention one of them, the *apostrophus*. The apostrophus notation is a system to represent multiplication by ten, starting with the numbers 500 and 1000. Roman numerals originally had their own symbols that were probably derived from tally marks. Gradually, these symbols were replaced by letters of the Roman alphabet that looked similar to them. The apostrophus notation evolved from the old symbol for 1000, which was written as ⅭⅮ (derived from the Greek Φ) or CIↃ. Adding a C on the left and the apostrophus, Ↄ, on the right, we obtain CCIↃↃ, which represents 10 times 1000, or 10,000. Adding another "pair of parentheses," we get CCCIↃↃↃ, which is 10 times 10,000 or 100,000, and so on. Splitting the symbol CIↃ into halves gives CI and IↃ, which turned into D for 500. Analogously, the symbols IↃↃ and IↃↃↃ would represent "halves," that is 5,000 and 50,000, respectively. In not strictly mathematical contexts, the symbol CIↃ could also represent, symbolically, some very large number, too large to be counted. It is conjectured that the English mathematician John Wallis (1616–1703) introduced the infinity symbol, ∞, which he had derived from the roman symbol CIↃ.[16]

As we said earlier, roman numerals are still everywhere around us; in almost every larger city, you will find them on building faces, cornerstones, or gravestones, especially in the older towns in Europe. Being able to "read" them can be both entertaining and useful. The fact that the roman numeral system doesn't use place values can also be an advantage. For young children, it is much easier to understand an additive numeral system than a place-value system. They will learn roman numerals very quickly and probably enjoy writing their age or their birthday in this style. Moreover, looking out for roman numbers on a museum tour or on a walk in the city and deciphering them will be fun and informative at the same time. Now that we have seen the development of representing numbers, we take that one step further to another counting aspect of our society, namely, the calendar.

MATHEMATICS ON THE CALENDAR

All too often, we take for granted how we got to the calendar that we use today, without any further thought about its accuracy, and how it became the standard of the world. It is therefore appropriate that we now trace the development of the present-day Gregorian calendar, and how it has changed over the years. Of course, we will consider the relationship of the calendar to astronomy, as time even in antiquity was measured by observing the motions of bodies that move in unchanging cycles. The only motions of this nature are those of the celestial bodies. Hence, we owe to astronomy the establishment of a secure basis for the measurement of time by determining the lengths of the day, the month, and the year. A year is defined as the interval of time between two passages of the earth through the same point in its orbit in relation to the sun. This is the solar year. It is approximately 365.24217 mean solar days. The length of the year is not commensurable with the length of the day; the history of the calendar is essentially the history of the attempts to adjust these incommensurable units in such a way as to obtain a simple and practical system.

The calendar story goes back to Romulus, the legendary founder of Rome, who introduced a year of 300 days divided into 10 months. His successor, Numa Pompilius (753 BCE–673 BCE), added 2 months to this calendar, which was used for the following six and a half centuries, until Julius Caesar (100 BCE–33 BCE) introduced the Julian calendar, which has a regular year of 365 days divided into 12 months. Every fourth year, a day is added to February, making the fourth year a leap year. Thus, the Julian year is on average 365.25 days long. The Julian calendar was widely spread abroad, along with other features of Roman culture, and it was generally used until 1582. If the solar year's length were indeed 365.25 days, then the introduction of an additional day once every 4 years would completely compensate for the discrepancy. The difficulty with this method of reckoning was that 365.25 was not the same as 365.24217, which is the actual time for the earth's revolution about the sun. Although it may seem an insignificant difference, in hundreds of years it would accumulate to a discrepancy of a considerable number of days. The Julian year was somewhat too long, and by 1582 the accumulated error amounted to 10 days.

Pope Gregory XIII tried to compensate for the error. Because the vernal equinox occurred on March 11 in 1582, he ordered that 10 days be suppressed from the calendar dates in that year so that the vernal equinox would fall on March 21, as it should. When he proclaimed the calendar reform, he formulated the rules regarding the leap years. The Gregorian calendar designates leap years as those that are divisible by 4, unless they are divisible by 100 and not 400. Thus, 1700, 1800, 1900, 2100, . . . are not leap years, while the year 2000 was a leap year, since it is divisible by 400. In the Gregorian calendar, 400 consecutive years (for example, from 1700 to 2099) contain 97 leap years, three less than in the Julian calendar. The mean length of a Gregorian year is therefore $\frac{400 \cdot 365 + 97}{400} = 365.2425$ days.

In Great Britain and its colonies, the change of the Julian to the Gregorian calendar was not made until 1752. In September of that year, 11 days were omitted. The day after September 2 was September 14. It is interesting, therefore, to see a copy of the calendar for September 1752 taken from the almanac of Richard Saunders—better known as *Poor Richard's Almanack* (see fig. 1.7).[17]

HISTORICAL HIGH POINTS   41

1752		September hath XIX Days this Year.				
		First Quarter, the 15th day at 2 afternoon. Full Moon, the 23rd day at 1 afternoon. Last Quarter, the 30th day at 2 afternoon.				
M D	W D	Saints' Days Terms, &c.	Moon South	Moon Sets	Full Sea at Lond.	Aspects and Weather
1	f	Day br. 3. 35	3 A 27	8 A 29	5 A 1	♊ ♃ ♀
2	g	London burn.	4 26	9 11	5 38	Lofty winds
		According to an act of Parliament passed in the 24th year of his Majesty's reign and in the year of our Lord 1751, the Old Style ceases here and the New takes its place; and consequently the next Day, which in the old account would have been the 3d is now to be called the 14th; so that all the intermediate nominal days from the 2d to the 14th are omitted or rather annihilated this Year; and the Month contains no more than 19 days, as the title at the head expresses.				
14	e	Clock slo. 5 m.	5 15	9 47	6 27	Holy Rood D. and hasty showers
15	f	Day 12 h. 30 m.	6 3	10 31	7 18	
16	g		6 57	11 23	8 16	
17	A	15 S. Aft. Trin.	7 37	12 19	9 7	More warm and dry weather
18	b		8 26	Morn.	10 22	
19	c	Nat. V. Mary	9 12	1 22	11 21	
20	d	Ember Week	9 59	2 24	Morn.	
21	e	St. Matthew	10 43	3 37	0 17	☌ ♀ ♀
22	f	Burchan	11 28	☾ rise	1 6	♊ ♃ ♀
23	g	Equal D. & N.	Morn.	6 A 13	1 52	☌ ☉ ☌
24	A	16 S. Aft. Trin.	0 16	6 37	2 39	☌ ☉
25	b		1 5	7 39	3 14	
26	c	Day 11 h. 52 m.	1 57	8 39	3 48	Rain or hail ☌ ☿ ♀ now abouts ✳ ♄ ♀
27	d	Ember Week	2 56	8 18	4 23	
28	e	Lambert bp.	3 47	9 3	5 6	
29	f	St. Michael	4 44	9 59	5 55	
30	g		5 43	11 2	6 58	

Figure 1.7.

Mathematicians have pondered the question of the calendar and tried to develop ways of determining the days of any given date or holiday. To develop a method for determining the day, we need to be aware that a calendar year (except for a leap year) is 52 weeks and one day long. If New Year's Day in some year following a leap year occurs on a Sunday, the next New Year's Day will occur on Monday. The following New Year's Day will occur on a Tuesday. The New Year's Day of the leap year will occur on a Wednesday. Since there are 366 days in a leap year, the next New Year's Day will occur on a Friday, and not on a Thursday. The regular sequence is interrupted every 4 years (except during years whose numbers are evenly divisible by 100 but not evenly divisible by 400).

Let's first try to develop a method to find the weekday for dates in the same year. Suppose February 4 falls on Monday. On which day of the week will September 15 fall? Assuming that this calendar year is not a leap year, one need only find the number of days between February 4 and

September 15. We first find that February 4 is the 35th day of the year and that September 15 is the 258th day of the year. (Table 1.5 expedites this.) The difference of 258 and 35, that is, 223, is the number of days between dates. Since there are 7 days in a week, divide 223 by 7, which yields $\frac{223}{7} = 31 + \frac{6}{7}$, or, stated another way, 31 with a remainder of 6. The 6 indicates that the day on which September 15 falls is the sixth day after Monday, which is Sunday. In the case of a leap year, one day must be added after February 28 to account for February 29.

Date	1	2	3	4	5	6	7	8	9	10	11	12	13	14	15	16
Jan.	1	2	3	4	5	6	7	8	9	10	11	12	13	14	15	16
Feb.	32	33	34	35	36	37	38	39	40	41	42	43	44	45	46	47
Mar.	60	61	62	63	64	65	66	67	68	69	70	71	72	73	74	75
Apr.	91	92	93	94	95	96	97	98	99	100	101	102	103	104	105	106
May	121	122	123	124	125	126	127	128	129	130	131	132	133	134	135	136
Jun.	152	153	154	155	156	157	158	159	160	161	162	163	164	165	166	167
Jul.	182	183	184	185	186	187	188	189	190	191	192	193	194	195	196	197
Aug.	213	214	215	216	217	218	219	220	221	222	223	224	225	226	227	228
Sept.	244	245	246	247	248	249	250	251	252	253	254	255	256	257	258	259
Oct.	274	275	276	277	278	279	280	281	282	283	284	285	286	287	288	289
Nov.	305	306	307	308	309	310	311	312	313	314	315	316	317	318	319	320
Dec.	335	336	337	338	339	340	341	342	343	344	345	346	347	348	348	350

Date	17	18	19	20	21	22	23	24	25	26	27	28	29	30	31
Jan.	17	18	19	20	21	22	23	24	25	26	27	28	29	30	31
Feb.	48	49	50	51	52	53	54	55	56	57	58	59			
Mar.	76	77	78	79	80	81	82	83	84	85	86	87	88	89	90
Apr.	107	108	109	110	111	112	113	114	115	116	117	118	119	120	
May	137	138	139	140	141	142	143	144	145	146	147	148	149	150	151
Jun.	168	169	170	171	172	173	174	175	176	177	178	179	180	181	
Jul.	198	199	200	201	202	203	204	205	206	207	208	209	210	211	212
Aug.	229	230	231	232	233	234	235	236	237	238	239	240	241	242	243
Sept.	260	261	262	263	264	265	266	267	268	269	270	271	272	273	
Oct.	290	291	292	293	294	295	296	297	298	299	300	301	302	303	304
Nov.	321	322	323	324	325	326	327	328	329	330	331	332	333	334	
Dec.	351	352	353	354	355	356	357	358	359	360	361	362	363	365	

Table 1.5.

HISTORICAL HIGH POINTS   43

A similar method for finding the weekday of the dates in the same year can be discussed as follows. Because January has 31 days, and $\frac{31}{7} = 4 + \frac{3}{7}$, the same date in the subsequent month will be 3 days after that day in January. The same date in March will also be 3 days later than in January. In April, it will be 6 days later than in January. We can then construct a table of index numbers, as shown in table 1.6, for the months that will adjust all dates to the corresponding dates in January:

January = 0	April = 6	July = 6	October = 0
February = 3	May = 1	August = 2	November = 3
March = 3	June = 4	September = 5	December = 5

Table 1.6.

The index numbers in table 1.6 provide the days between the months divided by 7 to get the excess days, as we did in the previous method. We need only add the date to the index number of the month and divide by 7, and then the remainder will indicate the day of the week. Let's consider an example. Suppose we use the year 1925. January 1 was on a Thursday, so we shall try to determine on which day of the week March 12 will fall. To do this, add $12 + 3 = 15$. Then divide that sum by 7: $\frac{15}{7} = 2$, with a remainder of 1. This indicates that it was on the same weekday as January 1, a Thursday. In leap years, an extra 1 has to be added for dates after February 29.

You may now want to find the day for a date for any given year. To do this, we first need to know on what day of the week January 1 of the year 1 fell, and also make adjustments for leap years. The day of the week on which January 1 of year 1 fell can be determined as follows. Using a known day and date, we find the number of days that have elapsed since January 1 of the year 1. Thus, since January 1, 1952, was on a Wednesday, in terms of the value of the Gregorian year, the number of days since January 1 is $1951 \cdot 365.2425 = 712588.1175$. Dividing by 7, we get 101,798, with a remainder of 2. The remainder indicates that 2 days should be counted from Wednesday (the day of the week on which Jan. 1, 1952 fell). Since calculations refer to the past, the counting is done backward, indicating that January 1 of year 1 (in the Gregorian calendar) fell on Monday.

One method for determining the day of the week for any year suggests that dates in each century be treated separately. If you know the day of the week of the first day of that period, you could, in the same fashion as before, determine the excess days after that day of the week (thus, the day of the week that a given day would fall on for that century). For the years 1900–1999, the information needed is:

1. The index numbers of the months (see table 1.6).
2. January 1 of 1900 was Monday.
3. The number of years (thus giving the number of days over the 52-week cycles) that have elapsed since the first day of the year 1900.
4. The number of leap years (i.e., accounting for additional days) that have occurred since the beginning of the century.

If we know this, we can ascertain how many days in that Monday-week cycle we need to count. To better understand this, consider the following example. We seek to determine on which day of the week May 9, 1914, fell. Begin by adding 9, which is a number of days in the month, to the number 1, which is the index number of the month, to the number of years, 14, and to the number 3, since there were three leap years to that date in that century. Therefore, we get $9 + 1 + 14 + 3 = 27$. We then divide by 7, which leaves a remainder of 6, and that indicates a Saturday.

As another example, let's try to determine on which day of the week August 16, 1937, fell. Here we begin by considering the following numbers for the sum: there were 16 days in the month up until that date; the index number for August is 2; 37 years have passed since the century began; and there were 9 leap years prior to that date. Thus, we get the sum $16 + 2 + 37 + 9 = 64$, which, when divided by 7, leaves a remainder of 1, which indicates a Monday.

For the period 1800–1899, the same procedure is followed, except that January 1, 1800, was on Wednesday. For the period September 14, 1752 (the day after the deduction of 11 days), through 1799, the same procedure is followed except that the first day of that period would be Friday. For dates before September 2, 1752, we do not have to treat dates in each century separately, since the Julian calendar applies, and thus the number of leap years that occurred until a certain date in any century is just the integral part of the number obtained by dividing the year by 4. Otherwise the same procedure is followed, except that the whole year would be added and the number of the days would start with Friday. You can see how if we apply our procedure to determine the day of the week that May 13, 1240, appeared, we would get the following sum: $13 + 1 + 1240 + 310 = 1564$. Then, dividing by 7 leaves a remainder of 3, indicating that May 13, 1240, was a Sunday.

There is another method for determining the day without having to consider separate periods. Again we start by knowing the day of January 1 of the year 1. We will not count the actual number of days that have elapsed since January 1 of year 1, but we will count the number of excess days over weeks that have elapsed and to this number we will add the number of days that have elapsed since January 1 of the given year. This total must be divided by 7, and the remainder will indicate the number of days that must be counted for that week; thus, we can establish a formula as follows:

1 (Monday) + the remainder of the division by 7 of (the number of years that have elapsed thus far + the number of days that have elapsed since January 1 of the given year + the number of leap years that have occurred since year 1) = the number of the day of the week.

46   THE MATHEMATICS OF EVERYDAY LIFE

The calculation of the number of leap years must take into account the fact that the years whose number ends with two zeros, and which are not divisible by 400, are not leap years. To better understand this, consider the following example: let's determine the day of the week on which the date December 25, 1954, fell. Here is a calculation: 1 + 1953 + 488 (leap years) − 15 (century wrongly assumed leap years 19 − 4) + 358 (number of days between January 1, 1954, and December 25, 1954) = 2785, which, when divided by 7, gives a remainder of 6. Thus, December 25, 1954, fell on the sixth day of the week: a Saturday.

Many other tables and mechanisms have been devised to solve the problem of determining days. One such, shown in figure 1.8, consists of four scales and is to be used as follows:

1. With a straightedge, join the point on the first scale indicating the date with the proper month on the third scale. Mark the point of intersection with the second scale.
2. Join this point on the second scale with the point on the fourth scale indicating the proper century. Mark the point of intersection with the third scale.
3. Join this point with the point indicating the appropriate year on the first scale. The point of intersection with the second scale gives the desired day of the week. (Note: For the months of January and February, use the year diminished by 1.)

HISTORICAL HIGH POINTS

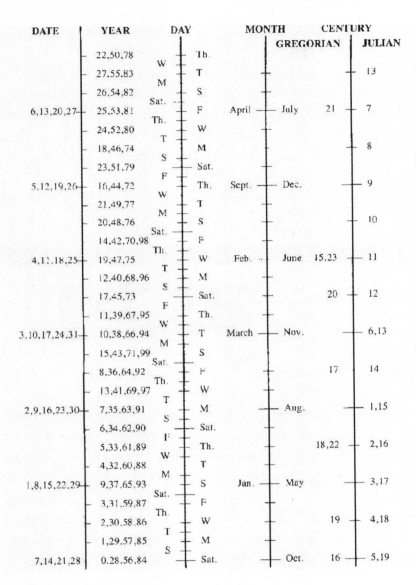

Figure 1.8.

The problem of a perpetual calendar occupied the attention of mathematicians, many of whom devoted considerable attention to calculating the date of Easter Sunday. We will address this in the next section.

48 THE MATHEMATICS OF EVERYDAY LIFE

HOW WE OVERLOOK OUR CALENDAR

Most of us are in daily contact with our calendar. However, there are certain peculiarities about the calendar that we ought to be aware of so that we can appreciate more than just looking up the day of the week. We really ought to know more about the calendar than the obvious. These little surprises can be used for entertainment purposes, or just to appreciate the construct of our calendar.

Consider any calendar page, say, October 2019. A randomly selected 3 · 3 square array of numbers is shown in table 1.7.

		1	2	3	4	5
6	7	8				12
13	14	15				19
20	21	22				26
27	28	29	30	31		

Table 1.7. October 2019.

Now focusing on the shaded dates, we will add 8 to the smallest number (9) in the shaded region and then multiply that sum by 9 to get $(9 + 8) \cdot 9 = 153$. If we now multiply the sum of the numbers of the middle row (which is 51) of this shaded matrix by 3, surprise! We get the same answer at which we previously arrived, 153. But why? Here are some clues: the middle number is the mean (or average) of the 9 shaded numbers, which is in itself a neat curiosity. The sum of the numbers in the middle column is $\frac{1}{3}$ of the sum of the nine numbers. An enthusiastic reader will likely want to justify these curiosities.

There's still more to admire about our calendar. What do you think the likelihood is that the following dates will all be on the same day of the week? The dates in question are: 4/4, 6/6, 8/8, 10/10, and 12/12 (where 4/4 represents April 4th, 6/6 represents June 6th, and so on).

HISTORICAL HIGH POINTS   49

Much to most people's surprise, these dates will always land on the very same day of the week, every year. Closer inspection will reveal that these dates are all exactly nine weeks apart. Such little-known facts always draw an interest that otherwise would be untapped. Counting on a calendar presents numerous surprises.

Suppose you find an old, unused calendar, and you would like to know when the calendar can be used again so that each date is on the correct day of the week for that particular year. For the twentieth and twenty-first century, the following procedure would work. To make matters simple, let's designate a particular year with the letter Y.

- If Y is a leap year, then the calendar can be reused in the year $Y + 28$.
- If Y is the year following a leap year, then the calendar can be reused in the years $Y + 6$, $Y + 17$, and $Y + 28$.
- If Y is the second year after a leap year, then the calendar can be reused in the years $Y + 11$, $Y + 17$, and $Y + 28$.
- If Y is the third year after a leap year, then the calendar can be reused in the years $Y + 11$, $Y + 22$, and $Y + 28$.

We will the leave the justification of most of these claims to you, but we will explain why any calendar can be reused after 28 years, if Y and $Y + 28$ belong to the twentieth or twenty-first century. Note that 2000 was not a leap year; hence, any consecutive 28 years between 1901 and 2099 contain 7 leap years and 21 common years. Since both numbers are multiples of 7, a timespan of exactly 28 years (between 1901 and 2099) will always correspond to an integer number of weeks. Therefore, all dates of the year will again fall on the same weekday after 28 years. Maybe now you won't take the calendar for granted as much as most people do in the course of normal daily use. As you can see, once again, mathematics can explain many things that we would hardly expect.

50   THE MATHEMATICS OF EVERYDAY LIFE

The second arrangement (see fig. 1.9) consists of three concentric rings intersected by seven radii. The procedure is:

1. Locate the date and the month on the outer ring; if they are two points, draw a line between them; if they coincide, draw a tangent.
2. Locate the century on the intermediate ring. Through this point draw a line parallel to the line drawn, until it intersects the intermediate ring at another point. The point found will be a ring-radius intersection.
3. From the point just found, follow the radius to the inner ring, then locate the year. (If the month is January or February, use the preceding year.) Draw a line between these two points on the inner ring. (If they coincide on Saturday, then Saturday is the weekday sought.)
4. Now find the point where the Saturday radius cuts the inner ring, and through this Saturday point draw a line parallel to the line just drawn. The line will meet the inner ring at some radius-ring intersection. The weekday on this latter radius is the weekday of the date with which we began.

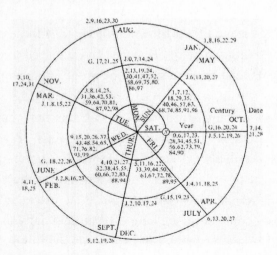

Figure 1.9.

HISTORICAL HIGH POINTS   51

As previously mentioned, many mathematicians spent a time analyzing the problem of a perpetual calendar and devoted considerable attention to calculating the date of Easter Sunday. All church holidays fall on a specific date. The ecclesiastical rule regarding Easter is, however, rather complicated. Easter must fall on the Sunday after the first full moon that occurs after the vernal equinox. Easter Sunday, therefore, is a movable feast that may fall as early as March 22 or as late as April 25. The following procedure to find Easter Sunday in any year from 1900 to 1999 is based on a method developed by the famous German mathematician Carl Friedrich Gauss (1777–1855).

1. Find the remainder when the year is divided by 4. Call this remainder a.
2. Find the remainder when the year is divided by 7. Call this remainder b.
3. Find the remainder when the year is divided by 19. Multiply this remainder by 19, add 24, and again find the remainder when the total is divided by 30. Call this remainder c.
4. Then add $2a + 4b + 6c + 3$. Divide this total by 7 and call the remainder d.

The sum of c and d will give the number of days after March 22 on which Easter Sunday will fall. As an example, let's try to determine what the date of Easter Sunday in 1921 was.

1. $\dfrac{21}{4} = 5 +$ remainder of 1

2. $\dfrac{21}{7} = 3 +$ remainder of 0

3. $\dfrac{21}{19} = 1 +$ remainder of 2; $\dfrac{2 \cdot 19 + 24}{30} = 2 +$ remainder of 2

4. $\dfrac{2 + 0 + 12 + 3}{7} = 2 +$ remainder of 3

5. $2 + 3 = 5$ days after March 22 is March 27.

52 THE MATHEMATICS OF EVERYDAY LIFE

(Note: The method above gives the date accurately except for the years 1954 and 1981. These years it gives a date exactly 1 week late, the correct Easters being April 18 and 19, respectively.)

We hope that this discussion gives you some insight into the complications of our calendar. For instance, we now understand why George Washington was born February 11, 1732, but we celebrate it on February 22. These curiosities await all who have a better understanding of the mathematical nature of our calendar, and now you perhaps won't take the calendar for granted!

Now with an insight into how numbers were developed over the past centuries and how we use them today, we are ready to embark on a journey of exploring the use of mathematics in our everyday life, where numbers still play an essential role.

CHAPTER 2

MATHEMATICS IN OUR EVERYDAY LIVES— ARITHMETIC SHORTCUTS AND THINKING MATHEMATICALLY

Almost every day we find ourselves in situations in which we apply, consciously or unconsciously, mathematical knowledge. Basic abilities in arithmetic are among those mathematical competencies we use most frequently. We may estimate the time we need to get from one place to another, combining different means of transportation, or calculating the total cost of the products in our shopping cart before we go to the register. Of course, one could argue that mastering arithmetic is nowadays superfluous and old-fashioned, since the cellphone practically everybody carries around is a smartphone, which can be used as a calculator. However, simple calculations can often be done much faster without the assistance of a calculator. It also keeps the brain fit, and using our own built-in "computer" is definitely more enjoyable than commanding an electronic device. But, most important, arithmetic offers an accessible playground for everyone to learn and develop mathematical thinking and problem-solving strategies, which can, in fact, be very helpful for making decisions in our everyday lives. We shall begin with some arithmetic shortcuts that can allow you to look at numbers and make some mental calculations—even faster than the calculator can.

ARITHMETIC WITH THE NUMBERS 9 AND 11

You may have wondered why certain numbers carry some special properties that allow a simplification of arithmetic processes, or, in other words, some arithmetic tricks. Let's consider some of these here. Both the number 11 and the number 9 are situated on either side of the number 10—which is the base of our number system—and therefore have very interesting properties. These properties give these numbers some unusual benefits in calculation. Let us begin by examining how we can determine divisibility by the numbers 9 and 11.

There are times in everyday-life situations when it is useful to know if a given number is divisible by 9 or by 3—especially if it can be done mentally! For example, suppose a restaurant bill of $71.22 needs to be split into three equal parts. Before actually doing the division, the thought about whether or not it is possible to split the bill equally into three parts may come into question. Wouldn't it be nice if there were some mental arithmetic shortcut for determining this? Well, here comes mathematics to the rescue. We are going to provide you with a rule to determine if a number is divisible by 3 or divisible by 9. The rule, simply stated, is:

If the sum of the digits of a number is divisible by 3 (or 9), then the original number is divisible by 3 (or 9).

Perhaps an example would be best to introduce this technique. Consider the number 276,357. Let's test it for divisibility by 3 (or 9). The sum of the digits is $2 + 7 + 6 + 3 + 5 + 7 = 30$, which is divisible by 3, but not by 9. Therefore, the original number (276,357) is divisible by 3, but not by 9. However, the number 14,688 is divisible by both 3 and 9, since the sum of its digits is $1 + 4 + 6 + 8 + 8 = 27$, which is divisible by both 3 and 9.

Just to make sure you are comfortable with this procedure, we will consider another example. Is the number 457,875 divisible by 3 or 9? The sum of the digits is $4 + 5 + 7 + 8 + 7 + 5 = 36$, which is divisible by 9 (and then, of course, by 3 as well), so the number 457,875 is divisible by 3 and by 9. If by some remote chance it is not immediately clear to you whether the sum of the digits is divisible by 3 or 9 (perhaps because it might still be too large a sum), then take the sum of the digits of this just-found sum and continue the process until you can visually make a determination of divisibility by 3 or 9.

Now that you are an expert at determining whether a number is divisible by 3 or 9, we can go back to the original question about the divisibility of the restaurant bill of $71.22. Can it be divided into three equal parts? Because $7 + 1 + 2 + 2 = 12$, and 12 is divisible by 3, then we can conclude that $71.22 is divisible by 3. (Notice that we need not be concerned with the decimal, since it is the number comprised of the digits with which we are concerned.)

In case you are interested as to why this rule actually works, here is a brief explanation using very simple algebra. Consider the base-10 number $N = ab,cde$, where the letters a, b, c, d, and e represent the digits and, therefore, the value of the number can be expressed as follows:

$N = 10^4 a + 10^3 b + 10^2 c + 10d + e = (9+1)^4 a + (9+1)^3 b + (9+1)^2 c + (9+1)d + e$. Gathering those multiples of 9, we get

$N = [9M + (1)^4]a + [9M + (1)^3]b + [9M + (1)^2]c + [9 + (1)]d + e$
(where $9M$ indicates a different multiple of 9 each time).

Factoring these multiples of 9, we get $N = 9M[a + b + c + d] + a + b + c + d + e$, which implies that the entire expression will be divisible by 9 when the sum of the digits $a + b + c + d + e$ is divisible by 9.

Let us now consider if there is an analogous special property for divisibility by 11. If you have a calculator at hand, the problem is easily solved. But that is not always the case. Besides, there is such a clever "rule" for testing for divisibility by 11—one that it is worth knowing just for its charm.

The rule is quite simple:

If the difference of the sums of the alternate digits is divisible by 11, then the original number is also divisible by 11.

This sounds a bit complicated, but it really isn't. Let us take this rule a piece at a time. "The sums of the alternate digits" means you begin at one end of the number, taking the first, third, and fifth, digit (and so on), and add them. Then you add the remaining (even-placed) digits. Subtract the two sums, and then inspect this resulting difference for divisibility by 11.

Perhaps it might be best to demonstrate this through an example. Suppose we test 768,614 for divisibility by 11. The sums of the alternate digits are: $7 + 8 + 1 = 16$ and $6 + 6 + 4 = 16$.

The difference of these two sums is $16 - 16 = 0$, which is divisible by 11. (Remember, $\frac{0}{11}=0$.) Therefore, we can conclude that 768,614 is divisible by 11.

Another example might be helpful to firm up an understanding of this procedure. To determine whether 918,082 is divisible by 11, we once again find the sums of the alternate digits: $9 + 8 + 8 = 25$ and $1 + 0 + 2 = 3$. The difference of the two sums is $25 - 3 = 22$, which is divisible by 11, and so the number 918,082 is divisible by 11. Here we have an example of a technique that not only can be helpful but also demonstrates the power and consistency of mathematics.

We would be remiss if we did not provide a justification for this rather-unexpected technique for determining whether a number is divisible by 11. Here is a brief discussion about why this rule works as it does. Consider the base-10 number $N = ab,cde$, where the letters a, b, c, d, and e represent the digits and, therefore, the value of the number can be expressed as

$N = 10^4a + 10^3b + 10^2c + 10d + e = (11-1)^4a + (11-1)^3b + (11-1)^2c + (11-1)d + e.$

If we let $11M$ represent a number which is a multiple of 11 (and it can be a different number each time, but still a multiple of 11), we can express the above equation as: $N = [11M + (-1)^4]a + [11M + (-1)^3]b + [11M + (-1)^2]c + [11 + (-1)]d + e$ or $N = 11M[a + b + c + d] + a - b + c - d + e$, which implies that divisibility by 11 of the number N is dependent on the divisibility of $a - b + c - d + e$, which written another way, is $(a + c + e) - (b + d)$, which is actually the difference of the sums of the alternate digits.

Having now considered rules for divisibility by these two special numbers, 9 and 11, let's consider other properties that these numbers have, to simplify our arithmetic processes. Perhaps one of the simplest mathematical tricks is to multiply by 11, mentally! This trick often gets a rise out of the unsuspecting mathematics-phobic person, because it is so simple that it is even easier than doing it on a calculator.

The rule is very simple:

To multiply a two-digit number by 11, just add the two digits and place this sum between the two digits.

Let's try using this technique. Suppose you need to multiply 45 by 11. According to the rule, add 4 and 5 and place this sum between the 4 and 5 to get 495.

This can become a bit more difficult when the sum of the two digits you are adding results in a two-digit number. We no longer have a single digit to place between the two original digits. So, if the sum of the two digits is greater than 9, then we place the units digit between the two digits of the number being multiplied by 11 and "carry" the tens digit to be added to the hundreds digit of the multiplicand. (Recall: The multiplicand is the number that is multiplied by another number, the multiplier—in this case, 11.) Let's try this procedure by finding the product of 78 · 11. We first get the sum of the digits: 7 + 8 = 15. We place the 5 between the 7 and 8, and then we add the 1 to the 7, to get [7 + 1][5][8], or 858.

It is fair to ask whether this technique also holds true when a number consisting of more than two digits is multiplied by 11. Let's go straight for a larger number such as 12,345 and multiply it by 11. Here we retain the first and last digit, then we begin at the right-side digit and add every pair of digits, going to the left: 1[1 + 2][2 + 3][3 + 4][4 + 5]5 = 135,795.

As was the case earlier, if the sum of two digits is greater than 9, then we place the units digit appropriately and carry the tens digit. To better understand how this is done, consider the following multiplication, 456,789 · 11:

Follow along as we carry the process step-by-step:

4[4 + 5][5 + 6][6 + 7][7 + 8][8 + 9]9
4[4 + 5][5 + 6][6 + 7][7 + 8][17]9
4[4 + 5][5 + 6][6 + 7][7 + 8 + *1*][7]9
4[4 + 5][5 + 6][6 + 7][16][7]9
4[4 + 5][5 + 6][6 + 7 + *1*][6][7]9
4[4 + 5][5 + 6][14][6][7]9
4[4 + 5][5 + 6 + *1*][4][6][7]9
4[4 + 5][12][4][6][7]9
4[4 + 5 + *1*][2][4][6][7]9
4[10][2][4][6][7]9
[4 + *1*][0][2][4][6][7]9
[5][0][2][4][6][7]9
5,024,679

This technique for multiplying by 11 might well be shared with your friends. Not only will they be impressed with your cleverness, they may also appreciate knowing this shortcut.

We now revert back to the number 9, as we search for a technique for multiplying any number by 9. Although this technique may be a bit cumbersome, especially when compared to using a calculator, this algorithm provides some insights into number theory, which is the basis for our understanding arithmetic processes. The number 9 has another unusual feature that enables us to use a surprising multiplication algorithm. Don't be distracted by the rather-complicated appearance. Just know that we present it here to indicate a multiplication property provided by the number 9. This procedure is intended for multiplying numbers of two digits or more by 9. It is best to discuss the procedure as we apply it to the multiplication 76,354 · 9.

Step 1	Subtract the units digit of the multiplicand from 10.	$10 - 4 = \mathbf{6}$
Step 2	Subtract each of the remaining digits (beginning with the tens digit) from 9, and then add this result to the previous digit in the multiplicand, read it from right to left. (For any two-digit sums, carry the tens digit to the next sum.)	$9 - 5 = 4, 4 + 4 = \mathbf{8}$ $9 - 3 = 6, 6 + 5 = 11, \mathbf{1}$ $9 - 6 = 3, 3 + 3 = 6, 6 + 1 = \mathbf{7}$ $9 - 7 = 2, 2 + 6 = \mathbf{8}$
Step 3	Subtract 1 from the left-most digit of the multiplicand.	$7 - 1 = \mathbf{6}$
Step 4	List the results in reverse order to get the desired product.	**687,186**

Table 2.1.

Although the technique here is not one that would be used to do the multiplication, we merely offer it for your amusement and so that the number 9 does not feel neglected in the multiplication process as compared to the number 11.

To compensate for the inability of the 9 to compete with the 11 in the process of multiplication, we shall present a procedure that the 9 provides beyond that of the 11.

HOW 9s CAN CHECK YOUR ARITHMETIC

As we mentioned earlier, the first occurrence in Western Europe of the Hindu-Arabic numerals we use today was in the book *Liber Abaci*, which was written in 1202 by Leonardo of Pisa (otherwise known as Fibonacci). As a young boy, Fibonacci traveled with his father, who directed a trading post in Bugia (in modern-day Algeria). Fibonacci traveled extensively along the Mediterranean coast, where he met merchants and became fascinated with the number system they used to do arithmetic. Recall that the publication of *Liber Abaci* provided the first use of these numerals in Europe. Before that, roman numerals were used extensively. They were, clearly, much more cumbersome to use for calculation than these ten numerals he had experienced in the Arabic world and which had originated in India.

Fascinated as Fibonacci was by the arithmetic calculations used in the Islamic world, in his book he introduced the system of "casting out nines"—which refers to a calculation check to determine if your result is possibly correct. The process requires subtracting a specific number of groups of 9 from the sum of the digits of the result (or, in other words, taking bundles of 9 away from the sum). Although this technique might come in handy, the nice thing about it is that it again demonstrates a hidden magic in ordinary arithmetic.

Before we discuss this arithmetic-checking procedure, we will consider how the remainder of a division by 9 compares to removing groups of 9 from the digit sum of the number. Let us find the remainder, when, say, 8,768 is divided by 9. Using a calculator, we find that the quotient is 974 with a remainder of 2.

This remainder can also be obtained by "casting out nines" from the digit sum of the number 8,768: This means that we will find the sum of the digits; and if the sum is more than a single digit, we shall repeat the procedure with the obtained sum. In the case of our given number, 8,768, the digit sum is 29 (8 + 7 + 6 + 8 = 29). Since this result is not a single-digit number, we will repeat the process with the number 29. Again, the casting-out-nines procedure is used to obtain 2 + 9 = 11; and, again, repeating this procedure for 11, we get 1 + 1 = 2, which is the same remainder as when we earlier divided 8,768 by 9.

We can now take this process of casting out nines to another application, that of checking multiplication. Perhaps it is best to see it applied. We would like to see if the following multiplication is correct: 734 · 879 = 645,186. We can check this by division, but that would be somewhat lengthy. We can also see if this product could be correct by casting out nines. To do that, we will take each of the factors and the product and then add the digits of each number—continuing this process as before until a single digit results:

For 734:   7 + 3 + 4 = 14; then 1 + 4 = 5.
For 879:   8 + 7 + 9 = 24; then 2 + 4 = 6.
For 645,186: 6 + 4 + 5 + 1 + 8 + 6 = 30; then 3 + 0 = 3.

The product of the first two factors is 5 · 6 = 30, which yields 3 by casting out nines (3 + 0 = 3); because this is the same as what we obtained when we added the digits of the product 645,186 (30), the answer could be correct.

For practice, we will do another casting-out-nines "check" for the following multiplication:

56,589 · 983,678 = 55,665,354,342
For 56,589:   5 + 6 + 5 + 8 + 9 = 33;   3 + 3 = 6
For 983,678:   9 + 8 + 3 + 6 + 7 + 8 = 41;   4 + 1 = 5
For 55,665,354,342: 5 + 5 + 6 + 6 + 5 + 3 + 5 + 4 + 3 + 4 + 2 = 48;
                     4 + 8 = 12;   1 + 2 = 3

To check for possibly having the correct product, we multiply the results from our first two factors: 6 · 5 = 30, or 3 + 0 = 3, which matches the 3 resulting from the product digit.

A similar procedure can be used to check for the likelihood of a correct sum or quotient, simply by taking the sum (or quotient) and casting out nines, then taking the sum (or quotient) of these "remainders" and comparing it with the remainder of the sum (or quotient). They should be equal if the answer is correct.

As we deal with the base-10 throughout our lives, it is good to see how the numbers on either side of the 10 have special properties because of their position relative to 10. Once again, because our culture has selected 10 as the base for our number system, we have had an opportunity to explore the peculiarities that evolve from this arrangement.

RULES FOR DIVISIBILITY

Having discussed some of the peculiarities of the numbers 9 and 11 (which included rules for divisibility), it is appropriate for us now to consider rules for divisibility by other numbers. We can easily determine when a number is divisible by 2 or by 5, simply by looking at the last digit (i.e., the units digit) of the number. That is, if the last digit is an even number (such as 2, 4, 6, 8, 0, and so on), then the number will be divisible by 2. A fun fact that is not as well known is that if the number formed by the last two digits of a given number is divisible by 4, then the number itself is divisible by 4. Also, if the number formed by the last three digits is divisible by 8, then the number itself is divisible by 8. You ought to be able to extend this rule to divisibility by higher powers of 2 as well.

Similarly, for 5, if the last digit of the number being inspected for divisibility is either a 0 or 5, then the number itself will be divisible by 5. If the number formed by the last two digits is divisible by 25, then the number itself is divisible by 25. The similarity of this rule to the previous one results from the fact that 2 and 5 are the prime factors of 10, which is the base of our decimal number system.

With the proliferation of the calculator, there is no longer a crying need to be able to detect by which numbers a given number is divisible. You can simply do the division on a calculator. Yet, for a better appreciation of mathematics, divisibility rules provide an interesting window into the nature of numbers and their properties. For this reason (among others), the topic of divisibility still finds a place on the mathematics-learning spectrum.

Most perplexing has always been to establish rules for divisibility by prime numbers (which are numbers whose only factors are 1 and the number itself). This is especially true of the rule for divisibility by 7, which follows a series of very nifty divisibility rules for the numbers 2 through 6. As you will soon see, the techniques for some of the divisibility rules for prime numbers are almost as cumbersome as an actual division algorithm, yet they are fun, and, believe it or not, they can come in handy. Let us consider the rule for divisibility by 7 and then, as we inspect it, let's see how this can be generalized for other prime numbers.

The rule for divisibility by 7: **Delete the last digit from the given number, and then subtract twice this deleted digit from the remaining number. If the result is divisible by 7, the original number is divisible by 7. This process may be repeated if the result is too large for simple inspection of divisibility of 7.**

Let's try an example to see how this rule works. Suppose we want to test the number 876,547 for divisibility by 7. Begin with 876,547 and delete its units digit, 7, and subtract its double, 14, from the remaining number: 87,654 − 14 = 87,640. Since we cannot yet visually inspect the resulting number for divisibility by 7 we continue the process.

We take the resulting number 87,640 and delete its units digit, 0, and subtract its double, still 0, from the remaining number; we get 8,764 – 0 = 8,764. This did not change the resulting number, 8,764, as we seek to check for divisibility by 7, so we continue the process.

Again, we take the resulting number 8,764 and delete its units digit, 4, and subtract its double, 8, from the remaining number; we get 876 – 8 = 868. Since we still cannot visually inspect the resulting number for divisibility by 7, we continue the process.

Continue with the resulting number 868 and delete its units digit, 8, and subtract its double, 16, from the remaining number. Doing this, we get 86 – 16 = 70, which is clearly divisible by 7. Therefore, the number 876,547 is divisible by 7.

Before we continue with our discussion of divisibility of prime numbers, you ought to practice this rule with a few randomly selected numbers and then check your results with a calculator.

Why does this rather-strange procedure actually work? The beauty of mathematics is that it clearly explains why some amazing procedures actually work. This will all make sense to you after you see what is happening with this procedure.

To justify the technique of determining divisibility by 7, consider the various possible terminal digits (that you are "dropping") and the corresponding subtraction that is actually being done by dropping the last digit. In the chart below you will see how dropping the terminal digit and doubling it to get the units digit of the number being subtracted gives us in each case a multiple of 7. That is, you have taken "bundles of 7" away from the original number. Therefore, if the remaining number is divisible by 7, then so is the original number, because you have separated the original number into two parts, each of which is divisible by 7, and, therefore, the entire number must be divisible by 7.

Terminal Digit	Number Subtracted from Original
1	$20 + 1 = 21 = 3 \cdot 7$
2	$40 + 2 = 42 = 6 \cdot 7$
3	$60 + 3 = 63 = 9 \cdot 7$
4	$80 + 4 = 84 = 12 \cdot 7$
5	$100 + 5 = 105 = 15 \cdot 7$
6	$120 + 6 = 126 = 18 \cdot 7$
7	$140 + 7 = 147 = 21 \cdot 7$
8	$160 + 8 = 168 = 24 \cdot 7$
9	$180 + 9 = 189 = 27 \cdot 7$

Table 2.2.

Now that we have a better understanding of why this works for divisibility by 7, let's examine the "trick" for divisibility by 13.

The rule for divisibility by 13: **This is similar to the rule for testing divisibility by 7, except that the 7 is replaced by 13 and, instead of subtracting twice the deleted digit, we subtract nine times the deleted digit each time.**

Let's check for divisibility by 13 for the number 5,616. Begin with 5,616 and delete its units digit, 6, then multiply it by 9 to get 54, which is then subtracted from the remaining number: $561 - 54 = 507$. Since we still cannot visually inspect the resulting number for divisibility by 13, we continue the process. Take the resulting number, 507, and delete its units digit and subtract nine times this digit from the remaining number: $50 - 63 = -13$. We see that -13 is divisible by 13, and, therefore, the original number is divisible by 13.

You might be wondering why we take the unit digit and multiply it by 9. To determine the "multiplier," 9 in this case, we sought the smallest multiple of 13 that ends in a 1. That was 91, where the tens digit is 9 times the units digit. Once again, consider the various possible terminal digits and the corresponding subtractions in the following table.

66 THE MATHEMATICS OF EVERYDAY LIFE

Terminal Digit	Number Subtracted from Original
1	90 + 1 = 91 = 7 · 13
2	180 + 2 = 182 = 14 · 13
3	270 + 3 = 273 = 21 · 13
4	360 + 4 = 364 = 28 · 13
5	450 + 5 = 455 = 35 · 13
6	540 + 6 = 546 = 42 · 13
7	630 + 7 = 637 = 49 · 13
8	720 + 8 = 728 = 56 · 13
9	810 + 9 = 819 = 63 · 13

Table 2.3.

In each case, a multiple of 13 is being subtracted one or more times from the original number. Hence, if the remaining number is divisible by 13, then the original number is divisible by 13. Let's move on to another prime number.

Divisibility by 17: **Delete the units digit and subtract five times the deleted digit each time from the remaining number until you reach a number small enough to determine its divisibility by 17.**

We justify the rule for divisibility by 17 as we did the rules for divisibility by 7 and 13. Each step of the procedure subtracts a "bundle of 17s" from the original number until we reduce the number to a manageable size and can make a visual inspection to determine divisibility by 17.

The patterns developed in the preceding three divisibility rules (for 7, 13, and 17) should lead you to develop similar rules for testing divisibility by larger primes. The following chart presents the "multipliers" of the deleted digits for various primes.

To Test Divisibility By ...	7	11	13	17	19	23	29	31	37	41	43	47
Multiplier	2	1	9	5	17	16	26	3	11	4	30	14

Table 2.4.

You may want to extend this chart. It's fun, and it will increase your perception of mathematics. You may also want to extend your knowledge of divisibility rules to include composite (i.e., non-prime) numbers. Why the following rule refers to relatively prime factors[1] and not just any factors is something that will sharpen your understanding of number properties. Perhaps the easiest response to this question is that relatively prime factors have independent divisibility rules, whereas other factors may not.

Divisibility by composite numbers: **A given number is divisible by a composite number if it is divisible by each of its relatively prime factors.** The chart below offers illustrations of this rule. You should complete the chart to 48.

To Be Divisible By ...	6	10	12	15	18	21	24	26	28
The Number Must Be Divisible By	2,3	2,5	3,4	3,5	2,9	3,7	3,8	2,13	4,7

Table 2.5.

At this juncture, you have not only a rather-comprehensive list of rules for testing divisibility but also an interesting insight into elementary number theory. Practice using these rules (to instill greater familiarity) and try to develop rules to test divisibility by other numbers in base-10 and to generalize these rules to other bases. Unfortunately, a lack of space prevents a more detailed development here. Yet we hope that these above examples have whet your appetite.

A QUICK METHOD TO MULTIPLY BY FACTORS OF POWERS OF 10

We all know that multiplying by powers of 10 is relatively easy. You need only place the appropriate number of zeros onto the number being multiplied by the power of ten. That is, 685 times 1,000 is 685,000. However, multiplying by factors of powers of 10 is just a bit more involved, but, in many cases, it can also be done mentally. Let's consider multiplying 16 by 25 (which is a factor of 100).

Since $25 = \frac{100}{4}$, $16 \cdot 25 = 16 \cdot \frac{100}{4} = \frac{16}{4} \cdot 100 = 4 \cdot 100 = 400$.

Nothing like a little practice to solidify a new algorithm. Following are a few more such examples:

$38 \cdot 25 = 38 \cdot \frac{100}{4} = \frac{38}{4} \cdot 100 = \frac{19}{2} \cdot 100 = 9.5 \cdot 100 = 950$

$1.7 \cdot 25 = \frac{17}{10} \cdot \frac{100}{4} = \frac{17}{4} \cdot \frac{100}{10} = 4.25 \cdot 10 = 42.5$.

In the previous line, in our last step we were able to break up the fraction $\frac{17}{4}$ as follows:

$\frac{17}{4} = \frac{16}{4} + \frac{1}{4} = 4 + 0.25 = 4.25$.

In an analogous fashion, we can multiply numbers by 125, since $125 = \frac{1000}{8}$.

Here are some examples of multiplication by 125 done mentally!

$32 \cdot 125 = 32 \cdot \frac{1000}{8} = \frac{32}{8} \cdot 1000 = 4 \cdot 1000 = 4000$

$78 \cdot 125 = 78 \cdot \frac{1000}{8} = \frac{78}{8} \cdot 1000 = \frac{39}{4} \cdot 1000 = 9.75 \cdot 1000 = 9750$

When multiplying by 50, you can use $50 = \frac{100}{2}$; when multiplying by 20, you can use $20 = \frac{100}{5}$.

Practice with these special numbers will clearly be helpful to you, since with some practice you will be able to do many calculations faster than the time it takes to find and then turn on your calculator!

ARITHMETIC WITH NUMBERS OF TERMINAL DIGIT 5

We offer now some other possibilities for mental calculation. Not that we want to detract you from using a calculator, but we wish merely to give you a sense of understanding and appreciation of number relationships that are part of our everyday lives.

Suppose we want to square 45. That is, $45^2 = 45 \cdot 45 = 2025$.

The process requires three steps:

Step 1	Multiply the multiples of 10 that are one higher and one lower than the number to be squared.	$40 \cdot 50 = 2000$
Step 2	Square the units digit 5.	$5 \cdot 5 = 25$
Step 3	Add the two results from steps 1 and 2.	$2000 + 25 = 2025$

Table 2.6.

In case you are curious why this works, we offer a short proof of this by using elementary algebra. Recall that the square of a binomial $(u + v)^2 = u^2 + 2uv + v^2$. We let $(10a + 5)$ be the multiple of 5 to be squared. Then we have $(10a + 5)^2 = 100a^2 + 100a + 25 = 100a(a + 1) + 25 = a(a + 1) \cdot 100 + 25$, which can be rewritten as $10a \cdot 10(a + 1) + 25$, which shows algebraically just what we did in the three steps above.

Here is an example of how to interpret this for an actual multiplication, say, $175 \cdot 175$:

Where $a = 17$, we get:

70 THE MATHEMATICS OF EVERYDAY LIFE

$x^2 = 10a \cdot 10(a+1) + 25 = 10 \cdot 17 \cdot 10(17+1) + 25 = \mathbf{170} \cdot \mathbf{180} + \mathbf{5} \cdot \mathbf{5} = 30600 + 25 = 30{,}625.$

Another way of looking at this technique is by considering the pattern that evolves below. Look at this list and see if you can identify the pattern of the two-digit numbers squared.

$05^2 = \mathbf{25}$
$15^2 = \mathbf{225}$
$25^2 = \mathbf{625}$
$35^2 = 1\mathbf{225}$
$45^2 = 20\mathbf{25}$
$55^2 = 30\mathbf{25}$
$65^2 = 42\mathbf{25}$
$75^2 = 56\mathbf{25}$
$85^2 = 72\mathbf{25}$
$95^2 = 90\mathbf{25}$

You will notice that in each case the square ends with 25 and the preceding digits are determined as follows:

$05^2 = \mathbf{0025},\ 0\ \ \ = 0 \cdot 1$
$15^2 = \mathbf{0225},\ 2\ \ \ = 1 \cdot 2$
$25^2 = \mathbf{0625},\ 6\ \ \ = 2 \cdot 3$
$35^2 = \mathbf{1225},\ 12\ = 3 \cdot 4$
$45^2 = \mathbf{2025},\ 20\ = 4 \cdot 5$
$55^2 = \mathbf{3025},\ 30\ = 5 \cdot 6$
$65^2 = \mathbf{4225},\ 42\ = 6 \cdot 7$
$75^2 = \mathbf{5625},\ 56\ = 7 \cdot 8$
$85^2 = \mathbf{7225},\ 72\ = 8 \cdot 9$
$95^2 = \mathbf{9025},\ 90\ = 9 \cdot 10$

The same rule can be extended to three-digit numbers and beyond. Take, for example, $235^2 = 55225$; the three digits preceding the last two digits (25), which are $552 = 23 \cdot 24$. The advantage of mental arithmetic tends to lose its attractiveness when we exceed two-digit numbers, since we must multiply two two-digit numbers—something usually not easily done mentally.

MULTIPLYING TWO-DIGIT NUMBERS LESS THAN 20

Aside from the electronic calculator, there are many methods for multiplying two-digit numbers up to 20. Here are some of these methods, which might also provide some insight to other methods.

Multiply $18 \cdot 17$ mentally. Here we seek to "extract" a multiple of 10 first—this time for the number $17 = 10 + 7$.

$18 \cdot 17 = 18 \cdot 10 + 18 \cdot 7 = 18 \cdot 10 + 10 \cdot 7 + 8 \cdot 7 = 180 + 70 + 56 = 306$

Or we can break up the number $18 = 10 + 8$, and work the method as follows:

$18 \cdot 17 = 10 \cdot 17 + 8 \cdot 17 = 10 \cdot 17 + 10 \cdot 8 + 8 \cdot 7 = 170 + 80 + 56 = 306$

Another method of multiplication would seek to obtain simpler factors—first for $18 = 20 - 2$:

$18 \cdot 17 = (20 - 2) \cdot 17 = 20 \cdot 17 - 2 \cdot 17 = 340 - 34 = 306$

Or, if we choose to use 17, $17 = 20 - 3$:

$18 \cdot 17 = 18 \cdot (20 - 3) = 18 \cdot 20 - 18 \cdot 3 = 360 - 54 = 306$

We can also use an entirely different method, albeit more of a novelty than a real help for quick multiplication, to multiply two numbers less than 20, such as 18 · 17, as follows:

Step 1	Select one of the two numbers you are multiplying (say, 18) and to it add the units digit of the other number (17).	18 + 7 = 25
Step 2	Place a zero at the end of this number.	250
Step 3	Multiply the two units digits of the two original numbers.	8 · 7 = 56
Step 4	Add the results of steps 2 and 3.	250 + 56 = 306

Table 2.7.

You might wish to try this technique with other two-digit numbers up to 20, probably more for entertainment than for practical use.

In the previous cases, we used the properties of binomial multiplication. Here we will use this in a more general way. Perhaps you can recall the following binomial multiplication: $(u + v) \cdot (u - v) = u^2 - uv + uv - v^2 = u^2 - v^2$, where u and v can take on any values that would be convenient to us. When we can apply this to the multiplication of 93 · 87, we notice that the two numbers are symmetrically distanced from 90. This allows us to do the following:

93 · 87 = (90 + 3) · (90 − 3) = $90^2 - 3^2$ = 8100 − 9 = 8091.

Here are a few further examples to help see the procedure in action.

42 · 38 = (40 + 2) · (40 − 2) = **$40^2 - 2^2$** = 1600 − 4 = 1596

64 · 56 = (60 + 4) · (60 − 4) = **$60^2 - 4^2$** = 3600 − 16 = 3584

As strange as this procedure may seem to be, with a little practice, it could come in quite handy when faced with this sort of multiplication problem.

Here are a few more examples to guide you along:

67 · 63 = (65 + 2) · (65 − 2) = **$65^2 - 2^2$** = [60 · 70 + 5 · 5] − 4 = 4225 − 4 = 4221

$26 \cdot 24 = (25 + 1) \cdot (25 - 1) = \mathbf{25^2 - 1^2} = [20 \cdot 30 + 5 \cdot 5] - 1 = 625 - 1 = 624$

MENTAL ARITHMETIC CAN BE MORE CHALLENGING—BUT USEFUL!

Here is another possible calculating shortcut that can be done mentally, of course, with some practice. Take, for example, the multiplication: $95 \cdot 97$.

Step 1	Add the numbers (95 + 97).	= 192
Step 2	Delete the hundreds digit.	= 92
Step 3	Add two zeros onto the number	= 9200
Step 4	$(100 - 95) \cdot (100 - 97) = 5 \cdot 3$	= 15
Step 5	Add the last two numbers	= 9215

Table 2.8.

This technique also works when seeking the product of two numbers that are farther apart. Let's examine $89 \cdot 73$:

$89 + 73 = 162$

$\mathbf{\bar{1}}62$ (Delete the hundreds digit.)

Add on two zeros = 6200

Then add $(100 - 89) \cdot (100 - 73) = 11 \cdot 27 = 297$ to get 6497.

For those who might be curious why this technique works, we can show you the simple algebraic justification:

We begin with the two-digit numbers represented by $100 - a$ and $100 - b$, where a and b are less than 100.

Step 1: $(100 - a) + (100 - b) = 200 - a - b$

Step 2: Delete the hundreds digit—which means subtracting 100 from the number:

$(200 - a - b) - 100 = 100 - a - b$

Step 3: Add on two zeros, which means multiplying by 100:

$(100 - a - b) \cdot 100 = 10,000 - 100a - 100b$

Step 4: $a \cdot b$

Step 5: Add the last two results:

$10,000 - 100a - 100b + a \cdot b$
$= 100 \cdot (100 - a) - (100b - ab)$
$= 100 \cdot (100 - a) - b \cdot (100 - a) = (100 - a) \cdot (100 - b)$, which is what we set out to show. Now you just need to practice this method to master it!

On paper you might think that some of the methods we are presenting as mental arithmetic are more complicated than doing the work with the traditional algorithms. Yet, with practice, some of these methods for two-digit numbers will become simpler to do mentally than writing them on paper and following the conventional steps.

ARITHMETIC WITH LOGICAL THINKING

In today's world, most people seem to neglect arithmetic shortcuts. With the ever-present electronic calculator, many people stop thinking arithmetically, and quantitatively. Yet there are times when we truly have to think, and neither the calculator nor any arithmetic shortcuts will help us find the required answer. We will begin with a little bit of logic that shows us the kind of thinking that we need to do in today's world.

There are times when simple arithmetic is not enough to answer a question. These are times when logical reasoning must be used to buttress the arithmetic. Let's consider one such example.

A woman passes her neighbor's house with her three sons, and her neighbor asks her how old her three sons are. She responds that, coincidentally, the product of their ages is 36, and the sum of their ages is the same number as his address. He looks puzzled as he stares at the house number, which is 13, and then he becomes even more puzzled when the woman tells him that she almost forgot one essential piece of information: her oldest son's name is Max. This really baffles him. How can this man determine the ages of the woman's sons? (We are only dealing with integer ages.)

To determine the ages of the three sons, the man considers which three numbers have a product of 36 and a sum of 13. They are (1, 6, 6) and (2, 2, 9).

So, when the woman tells the man that she left out an essential piece of information, it must have been to differentiate between the two sums of 13. When she mentioned that her oldest son is named Max, she was indicating that there is only *one* older son, thus eliminating the possibility of twins (or perhaps children of the same age having been born in the same year) of age 6. Hence, the ages of the three sons are 2, 2, and 9. Here arithmetic calculation alone did not help us answer the question; we needed to use some logical thinking. Let us now consider some clever methods for mental calculation.

USING THE FIBONACCI NUMBERS TO CONVERT KILOMETERS TO AND FROM MILES

When we say that mathematics is always available to help in your everyday life, we can demonstrate it very nicely by showing you a clever way to convert miles to and from kilometers. Most of the world uses kilometers to measure distance, while the United States still holds on to the mile to measure distance. This requires a conversion of units when one travels in a country in which the measure of distance is not the one to which we are accustomed. Such conversions can be done with specially designed calculators or by some "trick" method. That is where the famous Fibonacci numbers come in. Just to refresh your memory, the first few Fibonacci numbers are:

1, 1, 2, 3, 5, 8, 13, 21, 34, 55, 89, 144, . . .

Before we discuss the conversion process between these two units of measure, let's look at their origin. The mile derives its name from the Latin word for 1,000, *mille*, as it represented the distance that a Roman legion could march in 1,000 paces (which is 2,000 steps). One of these paces was about 5 feet, so the Roman mile was about 5,000 feet. The Romans marked off these miles with stones along the many roads they built in Europe—hence the name, "milestones"! The name *statute mile* (our usual measure of distance in the United States today) goes back to Queen Elizabeth I of England (1533–1603) who redefined the mile from 5,000 feet to 8 furlongs (5,280 feet) by statute in 1593. A furlong is a measure of distance within Imperial units and US customary units. Although its definition has varied historically, in modern terms it equals 660 feet, and is therefore equal to 201.168 meters. There are eight furlongs in a mile. The name "furlong" derives from the Old English words *furh* ("furrow") and *lang* ("long"). It originally referred to the length of the furrow in one acre of a ploughed open field (a medieval communal field that was divided into strips). The term is used today for distances horses run at a race track.

The metric system dates back to 1790, when the French National Assembly (during the French Revolution) requested that the French Academy of Sciences establish a standard of measure based on the decimal system, which they did. The unit of length they called a "meter" is derived from the Greek word *metron*, which means "measure." Its length was determined to be one ten-millionth ($1 \cdot 10^{-7}$) of the distance from the North Pole to the equator along a meridian going near Dunkirk, France, and Barcelona, Spain. Clearly, the metric system is better suited for scientific use than is the American system of measure. By an Act of Congress in 1866, it became "lawful throughout the United States of America to employ the weights and measures of the metric system in all contracts, dealings, or court proceedings."[2] Although it has not been used very often, there is curiously no such law establishing the use of our mile system.

To convert miles to and from kilometers, we need to see how one mile relates to the kilometer. The statute mile is exactly 1609.344 meters long. Translated into kilometers, this is 1.609344 kilometers. One the other hand, one kilometer is 0.621371192 miles long. The nature of these two numbers (reciprocals that differ by almost 1) might remind us of the golden ratio, which is approximately 1.618, and its reciprocal, which is approximately 0.618. Remember, it is the only number whose reciprocal differs from it by exactly 1. This would tell us that the Fibonacci numbers, the ratio of whose consecutive members approaches the golden ratio, might come into play here.

Let's see what length 5 miles would be in kilometers: 5 times $1.609344 = 8.04672 \approx 8$.

We could also check to see what the equivalent of 8 kilometers would be in miles: 8 times $0.621371192 = 4.970969536 \approx 5$. This allows us to conclude that approximately 5 miles is equal to 8 kilometers. Here we have two of our Fibonacci numbers.

As mentioned above, the ratio of a Fibonacci number to the one before it is approximately ϕ, which is the symbol used to note the golden ratio. Therefore, since the relationship between miles and kilometers is very close to the golden ratio, they appear to be almost in the relationship of consecutive Fibonacci numbers. Using this relationship, we would be able to approximately convert 13 kilometers to miles by replacing 13 with the previous Fibonacci number, 8. This would reveal to us that 13 kilometers is equivalent to about 8 miles. Similarly, 5 kilometers is about 3 miles, and 2 kilometers is roughly 1 mile. The higher Fibonacci numbers will give us a more accurate estimate, since the ratio of these larger consecutive Fibonacci numbers gets closer to ϕ.

Now suppose you want to convert 20 kilometers to miles. We have selected 20 because it is *not* a Fibonacci number. We can express 20 as a sum of Fibonacci numbers and convert each number separately and then add them. Thus, 20 kilometers = 13 kilometers + 5 kilometers + 2 kilometers. By replacing each of these Fibonacci numbers with the one lower, we have 13 replaced by 8, 5 replaced by 3, and 2 replaced by 1. This, therefore, reveals that 20 kilometers is approximately equal to $8 + 3 + 1 = 12$ miles. (Of course, if we would like to have a faster and perhaps less accurate estimate, we notice that 20 is close to the Fibonacci number 21. Using that number gives us 13 miles, a reasonable estimate done more quickly.)

Representing integers as sums of Fibonacci numbers is not a trivial matter. We can see that every natural number can be expressed as the sum of other Fibonacci numbers without repeating any one of them in the sum. Let's take the first few Fibonacci numbers to demonstrate this property as shown in table 2.9:

n	The Sum of Fibonacci Numbers Equal to n
1	1
2	2
3	3
4	1 + 3
5	5
6	1 + 5
7	2 + 5
8	8
9	1 + 8
10	2 + 8
11	3 + 8
12	1 + 3 + 8
13	13
14	1 + 13
15	2 + 13
16	3 + 13

Table 2.9.

You should begin to see patterns and also note that we used the fewest number of Fibonacci numbers in each sum in the table above. For example, we could also have represented 13 as the sum of $2 + 3 + 8$, or as $5 + 8$. Try to express larger natural numbers as the sum of Fibonacci numbers. Each time, ask yourself if you have used the fewest numbers in your sum.

To use this process to achieve the reverse, that is, to convert miles to kilometers, we write the number of miles as a sum of Fibonacci numbers and then replace each by the next *larger* Fibonacci number. Converting 20 miles to kilometers, therefore, gives us a sum as 20 miles = 13 miles + 5 miles + 2 miles. Now, replacing each of the Fibonacci numbers with their next larger in the sequence, we arrive at 20 miles = 21 kilometers + 8 kilometers + 3 kilometers = 32 kilometers.

To use this procedure, we are not restricted to use the Fibonacci representation of a number that uses the fewest Fibonacci numbers. You can use any combination of Fibonacci numbers whose sum is the number you are converting. For instance, 40 kilometers is $2 \cdot 20$, and we have just seen that 20 kilometers is 12 miles. Therefore, 40 kilometers is $2 \cdot 12 = 24$ miles (approximately). It should be noted that the larger the Fibonacci numbers being used, the more accurate the estimated conversion will be.

Consequently, we have another example of how some more sophisticated mathematics can be helpful in resolving a common, everyday problem.

THINKING "OUTSIDE THE BOX"

While grappling with a problem, very often we are asked by friends and colleagues to "think outside the box." Essentially, what is being suggested is to avoid trying to solve a problem in the traditional and expected fashion, and instead to look at the problem from a different

80 THE MATHEMATICS OF EVERYDAY LIFE

point of view. Practically by definition, this could be considered a counterintuitive way of thinking. This can even be true when a rather-simple problem is posed and the straightforward solution becomes a bit complicated. You could say that many people look at a problem in a psychologically traditional manner: the way it is presented and played out. To illustrate this point, we offer a problem here to convince you of an alternate method of thinking. Try the problem yourself (don't look below at the solution), and see whether you fall into the "majority-solvers" group. The solution offered later will probably enchant you, as well as provide future guidance to you.

The problem: **A single-elimination (one loss and the team is eliminated) basketball tournament has 25 teams competing. How many games must be played until there is a single tournament champion?**

Typically, the common way to approach this problem is to simulate the tournament, by selecting 24 teams to play in the first round (with one team drawing a bye). This will eliminate 12 teams (12 games have now been played). Similarly, of the remaining 13 teams, 6 play against another 6, leaving 7 teams in the tournament (18 games have been played now). In the next round, of the 7 remaining teams, 3 can be eliminated (21 games have so far been played). The four remaining teams play and eliminate 2 teams, leaving 2 teams for the championship game (23 games have now been played). This championship game is the 24th game.

A much simpler way to solve this problem, one that most people do not naturally come up with as a first attempt, is to focus only on the losers and not on the winners (as we have done above). We then ask the key question: "How many losers must there be in the tournament with 25 teams in order for there to be one winner?" The answer is simple: 24 losers. How many games must be played to get 24 losers? Naturally, 24. So there you have the answer, very simply done. Now most people will ask themselves, "Why didn't I think of that?" The answer is it was contrary to the type of training and experience we have had. Becoming aware of the strategy of looking at the problem from a different point of view may sometimes reap nice benefits, as was the case here.

Another way of looking at a solution to this problem—albeit quite similar to the previous solution—is to create an artificial situation where of the 25 teams, we will make 24 of these teams high-school-level players, and the 25th team a professional basketball team, such as the New York Knicks, which we will assume is superior to all of the other teams and will easily defeat each one. In this artificially construed situation, we would have each of the 24 high-school teams playing the Knicks, and, as expected, they would lose the game. Hence, after 24 games, a champion (in this case, the New York Knicks) is achieved.

In everyday life, we are often faced with simple issues that can be addressed by considering taking a look at the situation from an alternative point of view, just as we did in the above discussions. Suppose, for instance, you are asked to determine the number of people in attendance at a meeting of an association. Counting the members present would be unwieldy, since there may be many empty seats spread throughout the auditorium. Absentees all called in prior to the meeting to be excused. Therefore, you can solve the problem of determining the number present by subtracting the number of absentees from the total membership of the association. This exemplifies approaching the problem from a point of view different from simply counting or systematically "estimating" the attendance. If there are a small number of empty seats, counting the empty seats would be another way of determining the attendance (assuming you know what the total number of seats in the room is).

In any competitive sports event, the immediate tendency is to plan to use your strengths or strategies directly. An alternative point of view would be to assess and evaluate your competitor's strengths and weaknesses, and then generate your strategy from that assessment. Rather than viewing the impending contest and developing your game plan from your own vantage point, you could just as easily adopt a different point of view and assess it by a consideration of the competition.

Of course, this strategy has uses beyond the sports arena. It is interesting to note that in any form of negotiations, rather than only considering your own point of view, it is important to anticipate what position your "opponent" will take. Looking at the situation from this other point of view might help you find an appropriate direction for your own stance at these negotiations.

Another way of looking at this problem-solving strategy is to consider a detective investigating a case. She can sometimes select the guilty party from among several suspects not necessarily by proving that this one person committed the crime, but rather by adopting a different point of view and establishing that all of the other suspects had valid alibis, for instance. It is a process of elimination. Naturally, more substantial arguments would be necessary for a conviction, but at least this process would establish a direction for the investigation.

It is hoped that this logical demonstration will motivate you to consider alternative methods of solution even when faced with what appears to be a rather simple problem, and thereby establish a much more elegant method of solution.

SOLVING PROBLEMS BY CONSIDERING EXTREMES

Sometimes a clever solution to a simple mathematics problem reinforces a clever thinking procedure. One such strategy is sometimes used subconsciously by many people. We are referring to making a decision based on the process of using extremes. We often use extremes camouflaged in the phrase "the worst-case scenario." This approach might be considered as we decide which way to pursue an issue. Such a use of an extreme generally brings us to a good decision.

For instance, one can observe that the windshield of a car appears to get wetter the faster a car is moving in a rainstorm. This could lead some people to conclude that the car would not get as wet if it were to move slower. This leads to the natural next question that can be asked, namely, is it better to walk slowly or to run in a rainstorm so that you can minimize how much rain soaks your clothing? Setting aside the amount of wetness that the front of your body might get from the storm, let us consider two extreme cases for the top of the head: first, going infinitely fast, and, second, going so slowly as to practically be stationary. In the first case, there will be a certain amount of wetness on the top of the head. But, if we proceed at a speed of practically 0 mph, we would get drenched! Therefore, we conclude that the faster you move, the dryer you stay.

The previous illustration of this rather-useful problem-solving technique, *using extreme cases*, demonstrates how we use this strategy to clearly sort out an otherwise-cumbersome problem situation. We also use this same strategy of considering extremes in more everyday situations. A person who plans to buy an item where bargaining plays a part, such as in the process of buying a house or purchasing an item at a garage sale, must determine a strategy to make the seller an offer. He must decide what the lowest (extreme) price ought to be and what the highest (extreme) price might be, and then orient himself from there. In general, we often consider the extreme values of anything we plan to purchase and then make our decision about which price to settle on based on the extreme situations.

This is also relevant when it comes to scheduling. When you are budgeting time, you must consider extreme cases to be sure that time allocations are adequate. For example, allowing the maximum amount of time for each of a series of tasks would then enable an assurance of when the series of tasks would certainly be completed.

Extreme cases are also utilized when we seek to test a product, say, stereo speakers. We would want to test them at an extremely low volume and at an extremely high volume. We would then take for granted (with a modicum of justification), that speakers that pass the extreme-conditions test would also function properly between these extreme situations.

In mathematics, using extremes can be particularly useful. We often hear a problem that dramatically makes that point. A word about the problem before we present it. This is a problem that is very easy to understand. However, the beauty of the problem is the elegant solution that involves the use of considering an extreme. After you read the problem and consider a method of solution, allow yourself to perhaps struggle a bit before giving up (if you need to), and then consider the elegant solution provided here that is based on using an extreme situation. Here is the problem:

The problem: **We have two one-gallon bottles. One bottle contains a quart of red wine and the other bottle contains a quart of white wine. We take a tablespoonful of red wine and pour it into the white wine. Then we take a tablespoon of this new mixture (white wine and red wine) and pour it into the bottle of red wine. Is there more red wine in the white-wine bottle, or more white wine in the red-wine bottle?**

To solve the problem, we can figure this out in any of the usual ways—often referred to in the high-school context as "mixture problems"—or we can use some clever logical reasoning and look at the problem's solution as follows: With the first "transport" of wine, there is only red wine on the tablespoon. On the second "transport" of wine, there is as much white wine on the spoon as there remains red wine in the "white-wine bottle." This may require some abstract thinking, but you should "get it" soon.

The simplest solution to understand, and the one that demonstrates a very powerful strategy, is that of *using extremes*. Let us now employ this strategy for the above problem. To do this, we will consider the tablespoonful quantity to be a bit larger. Clearly the outcome of this problem is independent of the quantity transported. Therefore, let us use an *extremely* large quantity. We will let this quantity actually be the *entire* one quart—the extreme amount. Following the instructions given in the problem statement, we will take this entire amount (one quart of red wine), and pour it into the white-wine bottle. This mixture is now 50 percent white wine and 50 percent red wine. We then pour one quart of this mixture back into the red-wine bottle. The mixture is now the same in both bottles. Therefore, we can conclude that there is as much white wine in the red-wine bottle as there is red wine in the white-wine bottle, and the problem is solved!

We can consider another form of an extreme case, where the spoon doing the wine transporting has a zero quantity. In this case, the conclusion follows immediately: There is as much red wine in the white-wine bottle (none) as there is white wine in the red-wine bottle (none). Once again, by using extremes we very easily solved the problem in a rather-elegant fashion.

Another problem that can be rather easily solved by using an extreme condition is the following:

The problem: **A car is driving along a highway at a constant speed of 55 miles per hour. The driver notices a second car, exactly $\frac{1}{2}$ mile behind him. The second car passes the first, exactly 1 minute later. How fast was the second car traveling, assuming its speed is constant?**

Although this problem could be easily solved by using the traditional procedures taught in elementary algebra classes, it can be much more easily disposed of by considering an extreme situation. Assume that the first car is going *extremely* slowly, that is, at 0 mph. Under these conditions, the second car travels $\frac{1}{2}$ mile in one minute to catch the first car, which is to say that the second car would have to be traveling at a speed of 30 mph. Therefore, when the first car is moving at 0 mph, we find that the second car is traveling 30 mph faster than the first car. If, on the other hand, the first car is traveling at 55 mph (as was stated in the original problem), then the second car must be traveling at 30 + 55 = 85 mph (within the legal limit, we hope!).

We offer these problems merely as a demonstration of the power of thinking from the point of view of extremes, something that is used frequently in solving mathematical problems, as well as in the making of proper decisions in our everyday life.

THE WORKING-BACKWARD STRATEGY IN PROBLEM SOLVING

Often, without being directly aware that we are using a working-backward strategy, we find it to be a rather-useful approach. For example, the best approach to determine the most efficient route from one city to another depends on whether the starting point or the destination (endpoint) has numerous access roads. When there are fewer roads leading from the starting point, the forward method is usually superior. However, when there are many roads leading from the starting point and only one or two from the destination, an efficient way to plan the trip is to locate this final destination on a map, determine which roads lead most directly back toward the starting position, and then determine to which larger road that "last" access road leads. Progressively, by continuing in this way (i.e., *by working backward*), you land on a familiar road that is easily reachable from the starting point. At this step, you will have mapped out the trip in a very systematic way.

Let's consider a different example: A person who has an appointment in a distant city must determine the flight she will take to arrive comfortably on time for her meeting, yet not too far in advance. She begins by examining the airline schedule, starting with the arrival time closest to her appointment. Will she arrive in time? Is she "cutting it too close"? If so, she examines the next-earlier arrival time. Is this time all right? What if there is a weather delay? When is the next-earlier flight? Thus, by working backward, the traveler can decide the most appropriate flight to take to get to her appointment on time.

Let's examine another case of working backward. When a high-school freshman announces to a guidance counselor his desire to be admitted to a major Ivy League school and wants to know what courses to take, the counselor will usually look at what the potential college requires. At that point, the counselor begins to build the student's program for the next four years by working backward from the Advanced Placement courses that are usually taken in the senior year. However, to be prepared to take these courses, the student must first take some of the more-basic courses as a freshman, sophomore, and junior. For example, to take the Advanced Placement examination in calculus during senior year, the student must take the appropriate prerequisites. As we have stated previously, when there is a single final goal and we are interested in discovering the path to that endpoint from a starting point, we have a good opportunity to use the working-backward strategy.

The strategy game of Nim is another excellent example of when it is appropriate to use the working-backward strategy. In one version of the game, two players are faced with 32 toothpicks placed in a pile between them. Each player in turn takes 1, 2, or 3 toothpicks from the pile. The player who takes the final toothpick is the winner. Players develop a winning strategy by working backward from 32 (i.e., to win, the player must pick up the 28th toothpick, the 24th toothpick, etc.). Proceeding in this manner from the final goal of 32, we find that the player who wins picks the 28th, 24th, 20th, 16th, 12th, 8th, and 4th toothpicks. Thus, a winning strategy is to permit the opponent to go first and proceed as we have shown.

Although many problems may require some reverse reasoning (even if only to a minor extent), there are some problems whose solutions are dramatically facilitated by working backward. Consider the following problem:

The problem: **The sum of two numbers is 12, and the product of the same two numbers is 4. Find the sum of the reciprocals of the two numbers.**

The common approach is to immediately generate two equations $x + y = 12$ and $xy = 4$, where x and y represent the two numbers, and then to solve this pair of equations simultaneously by substitution. Assuming that the problem solver is aware of the quadratic formula (which is a staple topic in the high-school curriculum), the correct result will be a pair of rather unpleasant-looking values for x and y, namely, $x = 6 + 4\sqrt{2}$ and $y = 6 - 4\sqrt{2}$. Then we must find the reciprocals of these numbers and, finally, their sum. Can this problem be solved in this manner? Yes, of course! However, this rather-complicated solution process can be made much simpler by starting from the end of the problem, namely, with what we wish to find, $\frac{1}{x} + \frac{1}{y}$. One might logically ask, "What do we usually do when we see two fractions that are to be added?" If we compute the sum in the usual way, we obtain $\frac{x+y}{xy}$. However, since we were told at the outset that $x + y = 12$ and $xy = 4$, this fraction is a value $\frac{12}{4} = 3$, and the problem is solved! This is a dramatic example of how working backward trivializes a mathematical problem that done conventionally would be significantly more difficult.

Another everyday problem that requires working backward could be the following:

The problem: **Charles has an 11-liter can and a 5-liter can. How can he measure out exactly 7 liters of water?**

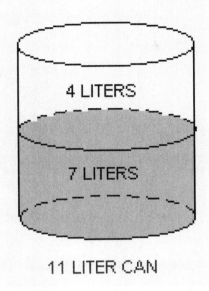

Figure 2.1.

Typically, a common procedure to resolve this problem is to keep "pouring" back and forth in an attempt to arrive at the correct answer, a sort of "unintelligent" guessing and testing. However, the problem can be solved in a more organized manner by using the working-backward strategy. To employ the strategy, we realize that we need to end up with 7 liters in the 11-liter can, leaving a total of 4 empty liters in the can. (See fig. 2.1) But where do 4 empty liters come from?

To obtain 4 liters, we must leave 1 liter in the 5-liter can. Now, how can we obtain 1 liter in the 5-liter can? Fill the 11-liter can and pour from it twice into the 5-liter can (fig. 2.2), discarding the water. This leaves 1 liter in the 11-liter can. Pour the 1 liter into the 5-liter can.

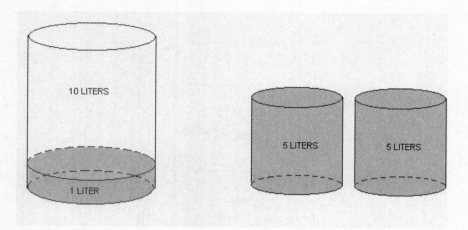

Figure 2.2.

Then fill the 11-liter can and pour off the 4 liters needed to fill the 5-liter can. (See fig. 2.3.) This leaves the required 7 liters in the 11-liter can. And the problem is solved!

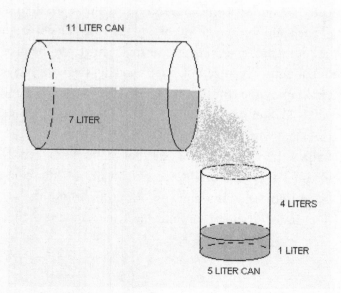

Figure 2.3.

Note that problems of this sort do not always have a solution. That is, if you wish to construct additional problems of this type, you must bear in mind that a solution exists only when the difference of multiples of the capacities of the two given cans can be made equal to the desired quantity. In this problem, $2 \cdot 11 - 3 \cdot 5 = 7$.

This concept can lead to a discussion of parity. We know that the sum of two like parities will always be even (i.e., even + even = even, and odd + odd = even), whereas the sum of two unlike parities will always be odd, odd + even = odd. Thus, if two even quantities are given, they can never yield an odd quantity. For example, given a 10-liter can and a 2-liter can, it is not possible to measure out an odd quantity of liters, since $n \cdot 10 - m \cdot 2 = (5n - m) \cdot 2$, which is even for all pairs of integers n and m. On the other hand, if at least one of the two cans holds an odd quantity of liters, we can measure out both odd and even quantities of liters (but not arbitrary ones).

Now that we have explored arithmetic shortcuts and clever methods of thinking logically, we are ready to apply our skills to everyday-life encounters.

CHAPTER 3

MATHEMATICAL APPEARANCES AND APPLICATIONS IN EVERYDAY-LIFE PROBLEMS

There is hardly a moment in our daily travels through our regular experiences when, in some form or another, mathematics does not present itself. In this chapter, we will expose some of these experiences as well as make you aware of how knowledge of some simple mathematics can facilitate your daily experiences and help you solve everyday life problems.

SHOPPING WITH MATHEMATICAL SUPPORT

Most supermarkets today provide the unit cost of an item. This is very helpful in that it allows the consumer to decide whether it makes sense to buy two 12 oz. jars of mayonnaise costing $1.35 per jar, or one 30 oz. jar of the same brand of mayonnaise costing $3.49. We have been trained to think that the larger quantity is generally the better price value. However, there is a neat little trick to determining which option offers the better price per ounce.

First we need to establish the price per ounce for each of the two different-sized jars:

94   THE MATHEMATICS OF EVERYDAY LIFE

For the 12-ounce jars, the price per ounce is: $\frac{\$1.35}{12}$.

For the 30-ounce jar, the price per ounce is: $\frac{\$3.49}{30}$.

To compare the two fractions, $\frac{1.35}{12}$? $\frac{3.49}{30}$, in order to see which is larger, we can implement a neat little algorithm to accomplish this task. We will cross multiply, writing the products under the fraction whose numerator was used. (See table 3.1)

Fraction to Determine Price Per Ounce	$\frac{1.35}{12}$	$\frac{3.49}{30}$
Cross Multiplication	1.35 · 30	3.49 · 12
Product of Cross Multiplication	40.50	41.88

Table 3.1.

The larger product (in this case, 41.88), determines which fraction (in this case, $\frac{3.49}{30}$) is larger, and therefore which is the more expensive. As we can see in this example, counterintuitively, the larger jar was more expensive per ounce than the two smaller jars. Although this is not typically expected, it does occur, and for that reason a wise consumer will make these comparisons.

SUCCESSIVE PERCENTAGES

As mentioned above, we encounter mathematical challenges in our everyday lives and don't even realize that they can be properly understood with just a little mathematics know-how. We often visit stores that are running a sale and then on a special day will *add* a percentage on top of the one that was previously advertised. The typical response is to add the two percentages and conclude that the total savings for

MATHEMATICAL APPEARANCES AND APPLICATIONS   95

the day would be the sum of the two percentages. Upon close reflection, you will see that this is clearly a wrong calculation. Most folks defer thinking about percentage problems, as they see them as nothing but a nemesis. Problems get particularly unpleasant when multiple percentages need to be processed in the same problem. However, we shall see how such successive percentages lend themselves very nicely to a delightfully simple arithmetic algorithm that leads us to lots of useful applications and provides new insight into successive percentage problems. We think you'll find this not-very-well-known procedure enchanting.

Let's begin by considering the following problem:

The problem: **Wanting to buy a coat, Barbara is faced with a dilemma. Two competing stores next to each other carry the same brand of coat with the same list price, but with two different discount offers. Store A offers a 10 percent discount year-round on all its goods, but, on this particular day, it is offering an additional 20 percent discount on top of its already-discounted price. Store B simply offers a discount of 30 percent on that day in order to stay competitive. Are the two end prices the same? If not, which gives Barbara the better price?**

At first glance, you may assume there is no difference in price, since $10 + 20 = 30$, which would appear to be yielding the same discount in both cases. Yet, with a little more thought, you may realize that this is not correct, since in store A only 10 percent is calculated on the original list price, while the 20 percent discount is calculated on the lower price (that is, the 10 percent discounted price); in comparison, at store B, the entire 30 percent is calculated on the original price. Now, the question to be answered is, What percentage difference is there between the discount in store A and store B?

To determine the difference in the prices, one procedure might be to assume the cost of the coat to be $100, and then calculate the 10 percent discount, yielding a $90 price, and then calculate an additional 20 percent of the $90 price (or $18), which will bring the price down to $72. In store B, the 30 percent discount on $100 would bring the price down to $70, giving a discount difference of $2 between the two stores, which in this case would be a 2 percent difference. This procedure, although correct and not too difficult, is a bit cumbersome and does not always allow a full insight into the situation, as you will soon see.

We shall provide an interesting and quite unusual procedure for a deeper look at this situation as well as for entertainment.

We will consider a somewhat mechanical method for obtaining a single percentage discount (or increase) equivalent to two (or more) successive discounts (or increases). Follow this four-step procedure:

(1) Change each of the percentages involved into decimal form:
   0.20 and 0.10
(2) Subtract each of these decimals from 1.00 if you are calculating a discount or decrease (for an increase, add to 1.00):
   0.80 and 0.90
(3) Multiply these decimals:
   $(0.80)(0.90) = 0.72$
(4) Subtract this number from 1.00:
   $1.00 - 0.72 = 0.28$, which, written as a percent, is 28 percent.
   This represents the combined *discount*.
   (If the result of step 3 is greater than 1.00, subtract 1.00 from it to obtain the percent of *increase*.)

When we convert .28 back to percent form, we obtain 28 percent, which is the equivalent of successive discounts of 20 percent and 10 percent.

Therefore, we can conclude that the combined percentage of

28 percent differs from the single discount of 30 percent by 2 percent. As such, in our example above, Barbara should purchase her coat from Store B.

Following the same procedure as above, you can also combine more than two successive discounts. Furthermore, successive increases, combined or not combined with a discount, can also be accommodated in this procedure by adding the decimal equivalent of the increase to 1.00 (recall, the discount was *subtracted* from 1.00), and then continue with the procedure in the same way. If the end result comes out greater than 1.00, then this will have resulted in an overall increase rather than the discount as found in the above problem.

A conundrum often facing consumers is that of determining whether a discount and increase of the same percentage leaves the original price unchanged. For example, suppose a store just increased all of its prices by 10 percent and then notices that its business has declined substantially, whereupon they then resort discounting all of these recently increased prices by the same percentage: 10 percent. Have they then restored the prices to their original level? Using this technique, we find ourselves multiply 1.1 times 0.90 to get 0.99, which would indicate that the original price had dropped by 1 percent. For many people this is a counterintuitive result. (Again, with further reflection, you'll see that the difference is a result of considering the original price prior to the 10 percent increase and then the new price prior to its 10 percent decrease. Ten percent of the lower price amounts to less than 10 percent of the higher price. By taking 10 percent off of the higher price, the store is therefore providing a steeper discount than it had for the increase.)

As you can see, this procedure not only streamlines a typically cumbersome situation but also provides some insight into the overall picture. For example, consider the following question: Is it advantageous to the buyer in the above problem, Barbara, to receive a 20 percent discount and then a 10 percent discount, or the reverse, a 10 percent discount and then a 20 percent discount? The answer to this question is not immediately intuitively obvious. Yet, since the procedure just presented shows that the calculation is merely multiplication, a commutative operation, we can immediately conclude that there is no difference between the two.

So here you have a delightful algorithm for combining successive discounts or increases or combinations of these to calculate the combined result. Not only is it useful, but also it gives you some newfound power in dealing with percentages when a calculator might not be available.

Another shopping situation in which mathematics can be helpful is when there are discounts of different types. Suppose you have two sales-promotion coupons for the same store, one that says "20 percent off" (independent of the purchase) and one that applies only if a certain minimal amount of money is spent, for example, "$15 off for purchases exceeding $49.99." Assuming that the two coupons cannot be combined, then which one would be more advantageous to use, if the item we want to buy costs, say, $80? The 20 percent off coupon would yield a price of $64, while the $15 reduction would yield a price of $65. It might be nice to know at which price the coupon for 20 percent off will become the more advantageous option.

To approach this problem, we may consider two extreme cases. Let's consider an item costing $50, since the minimal purchase that qualifies for the second coupon is $50, which when reduced by $15 would be $35. On the other hand, the fixed-percentage coupon would yield a reduction of 20 percent, or one fifth, of $50, which is $10. That means we would have to pay $40, if we used the fixed-percentage coupon. Thus, the $15-off coupon would be the better choice for a $50 purchase.

The other extreme case would be infinity, but since we can spend only a finite amount of money, let's assume our budget is very high, say it is $150. One fifth of $150 is $30. Therefore, the 20 percent coupon yields a reduced price of $120, while the other coupon (a reduction of $15) would result in a price of $135. Obviously, the better choice of coupon depends on the total sum of the purchase. For the fixed-percentage coupon, the amount of money saved increases with the price, whereas we cannot save more than $15 with the other coupon. The extreme cases we have considered show that there must be some break-even point X between $50 and $150 at which price both coupons yield the same discount. For purchases at $50, the 20 percent reduction will be $10 and the $15-off coupon is therefore preferred; but, for purchases at or above $50 and below X, the $15-off coupon is still the better choice. To find the break-even price X for a 20 percent coupon and a $15-off coupon, we just have to compute: $X - 0.20X = X - \$15$, and then $X = \$75$. Therefore, if we want to buy an item for more than $75, we should use the 20 percent coupon.

Occasionally you may also encounter different types of coupons that are combinable, although this is a rare phenomenon, since most stores are usually not that generous to their customers. Let's take a look at such a situation, since it provides an example for the mathematical notion of noncommutativity. Suppose we were allowed to use both coupons for the same purchase, that is, coupon A with a 20 percent discount as well as coupon B with a $15 reduction. Now the question arises whether or not the order of the discounts matters. If it were to matter, which of the two coupons should be used first? Denoting the price without any discount by P (which we assume to be at least $50 for the sake of simplicity), we obtain the following:

- a reduced price by using coupon A first is $p_{A,B} = P \cdot 0.8 - \$15$, and
- a reduced price by using coupon B first is $p_{B,A} = (P - \$15) \cdot 0.8$

To summarize, applying coupon B first gives us $p_{B,A} = (P - \$15) \cdot 0.8 = P \cdot 0.8 - \$12$; because that is more than $p_{A,B}$, which is $P \cdot 0.8 - \$15$, we should apply coupon A first, unless $P \cdot 0.8$ is less than $50. In this case, the $15-off coupon (B) would not be applicable anymore and we would have to use coupon B first to get the more favorable price.

100 THE MATHEMATICS OF EVERYDAY LIFE

Two operations that will in general lead to different results if their order is reversed are called "noncommutative" in mathematics. As our analysis showed, these different types of discounts are an example of noncommutative operations, meaning that the order does matter! You should think about that if you are offered combinable discounts of different types. However, requesting your preferred order might be a bit of a challenge.

RAISING INTEREST!

We are often confronted with advertisements by savings institutions offering attractive interest rates and frequent compounding of interest on deposits. Since most banks have a variety of programs, it's valuable for potential depositors to understand how interest is calculated under each of the available options. In our discussion of interest rates and practices we will use the formula for compound interest to calculate the return on investments at any rate of interest, for any period of time, and for any commonly used frequency of compounding, including instantaneous (continuous) compounding. These aspects will also determine which of two or more alternatives gives the best return over the same time period.

Let's consider the following interesting problem:

The problem: **In the year 1626, Peter Minuit bought Manhattan Island for the Dutch West India Company from the Native Americans called the Lenape (Indians) for trinkets costing 60 Dutch guilders, or about $24. Suppose the Lenape had been able to invest this $24 at that time at an annual interest rate of 6 percent, and suppose further that this same interest rate had continued in effect all these years. How much money could the present-day descendants of these the Lenape collect if (1) only simple interest were calculated, and (2) interest were compounded (a) annually, (b) quarterly, and (c) continuously?**

MATHEMATICAL APPEARANCES AND APPLICATIONS 101

The answers to these questions should surprise everyone!

Perhaps you will recall that simple interest is calculated by taking the product of the principal, P; the annual interest rate, r; and the time in years, t. Accordingly, you have the formula $I = Prt$, and in the above problem $I = (24)(0.06)(392) = \$564.48$ as simple interest. Add this to the principal of $\$24.00$ to obtain the amount A of $\$588.48$ available in 2018. You have just used the formula for "amount," $A = P + Prt$.

With this relatively small sum in mind (for a return after 392 years!), let's investigate the extent to which this return would have been improved if interest had been compounded annually instead of being calculated on only a simple basis. With a principal P, an annual rate of interest r, and a time $t = 1$, the amount A at the end of the first year is given by the formula $A_1 = P + Pr = P(1 + r)$. (The subscript indicates the year at the end of which interest is calculated.) Now $A_1 = P(1 + r)$ becomes the principal at the beginning of the second year, upon which interest will be credited during the second year.

Therefore, $A_2 = P(1+r) + P(1+r)r = P(1+r)(1+r) = P(1+r)^2$. Since the last expression represents the principal at the beginning of the third year, you have $A_3 = P(1+r)^2 + P(1+r)^2 r = P(1+r)^2(1+r) = P(1+r)^3$. By now, you will see the emerging pattern and should be able to suggest the generalization for the amount after t years,

$$A_t = P(1+r)^t.$$

Now try this formula on the $24 investment made in 1626! Assuming annual compounding at 6 percent per annum, you have $A_{392} = 24(1 + .06)^{392} = 199{,}576{,}970{,}308$. This means that the original $24 is now worth almost $200 billion! Can you believe the huge difference between this figure and the figure $588.48 obtained by computing simple interest?

Most banks now compound not annually but quarterly, monthly, daily, or continuously, so we shall next generalize the formula

$A = P(1 + r)^t$ to take into account compounding at more frequent intervals. Bear in mind that if interest is compounded semiannually, the periodic rate would be only one-half the annual rate, but the number of periods would be twice the number of years; so

$$A = P\left(1+\frac{r}{2}\right)^{2t}.$$

Likewise, if the interest is compounded quarterly, we have

$$A = P\left(1+\frac{r}{4}\right)^{4t}.$$

In general, if the interest is compounded n times a year, the formula would be

$$A = P\left(1+\frac{r}{n}\right)^{nt},$$

which can be used for any finite value of n. Letting $n = 4$ in our original problem yields

$$A = 24\left(1+\frac{.06}{4}\right)^{4(392)} = 24(1.015)^{2568} = 330{,}343{,}289{,}050.$$

The $24 has now risen to about $330 billion! Notice that changing the compounding from annually to quarterly increased the yield by about $130 billion. You may now wonder whether the yield can be increased indefinitely by simply increasing the frequency of compounding. A complete treatment of this question requires a thorough development of the concept of limits, but an informal, intuitive approach will suffice here. We shall first explore the simpler problem of an investment of $1 at a nominal annual interest rate of 100 percent for a period of one year. This will lead us to

$$A = 1\left(1+\frac{100}{n}\right)^n.$$

MATHEMATICAL APPEARANCES AND APPLICATIONS   103

You should now prepare a table of values for A for various common values of n, such as $n = 1$ (annual compounding), $n = 2$ (semiannual), $n = 4$ (quarterly), and $n = 12$ (monthly). Notice that the amount A does *not* rise astronomically as n increases, but rather rises slowly from $2.00 ($n = 1$) to about $2.60 ($n = 12$). The amount A would approach, but not quite reach, the value $2.72. Explaining this limiting value takes us a bit out of the realm of this book. (Yet, for the more advanced reader, it is

$$\lim_{n\to\infty}\left(1+\frac{1}{n}\right)^n = e = 2.71828\ldots.)$$

Since investments generally do not earn 100 percent interest, we must convert to a general interest rate, say, r. By letting $\frac{r}{n}=\frac{1}{k}$, we have $n = kr$ and

$$A = P\left(1+\frac{r}{n}\right)^{nt}, \text{ which then becomes}$$

$$A = P\left(1+\frac{1}{k}\right)^{krt} = P\left[\left(1+\frac{1}{k}\right)^k\right]^{rt}.$$

Clearly, as n approaches infinity, so does k, since r is finite, so the expression in brackets approaches the value e as a limit (e is the base of natural logarithms and has a value of approximately 2.718281828 4590452353602874713527 . . .). You then have the formula $A = Pe^{rt}$ for instantaneous compounding, where r is the nominal annual rate of interest and t is the time in years. You might be interested in knowing that this formula is a special representation of the general law of growth, which is usually written in the form $N = N_0 e^{rt}$, where N represents the final amount of a material whose initial amount was N_0. This law has applications in many other areas such as population growth (of people, of bacteria in a culture, etc.) and the radioactive decay of elements, in which case it becomes the law of decay, as $N = N_0 e^{-rt}$.

Completing the investment problem, using 2.72 as an approximation to e, you have $A = 24(2.72)^{.06(392)} = 399{,}274{,}766{,}704$. So you can see that the "ultimate" return on a $24 investment (at a nominal annual interest rate of 6 percent for 392 years) is almost $400 billion!

Banks currently offer much lower interest rates, and compounding is commonly done quarterly, monthly, daily, or continuously. You can work problems with varying principals, periodic rates, frequencies of compounding, and time periods, and compare yields. You may be surprised at the outcomes!

THE RULE OF 72

We often want to know how a certain interest rate in a bank will affect our total holdings. Naturally there are traditional ways of calculating interest, which we have just experienced in the previous section. However, there is an unusual quirk of our number system that allows us to calculate how long it will take to double your money in a bank with a daily compounding procedure at any given annual percentage rate. The procedure is as fast as you can divide 72 by another number. This is clearly good to know, but it is the unusualness of this rule that compels us to exhibit it here. It is called the "Rule of 72," as it is based on this number. Let's explore this further.

The "Rule of 72" states that, roughly speaking, money will double in $\frac{72}{r}$ years when it is invested at an annual compounded interest rate of r percent. So, for example, if we invest money at an 8 percent compounded annual interest rate, it will double its value in $\frac{72}{8} = 9$ years. Similarly, if we leave our money in the bank at a compounded rate of 2 percent, it would take 36 years for this sum to double its value. You might want to better understand why this is so, and how accurate it really is. The following discussion will explain that.

To investigate why or if this really works, we consider the compound-interest formula

$$A = P\left(1 + \frac{r}{100}\right)^n,$$

MATHEMATICAL APPEARANCES AND APPLICATIONS 105

where A is the resulting amount of money and P is the principal invested for n interest periods at r percent annually.

We need to investigate what happens when $A = 2P$.

The above equation then becomes $2 = \left(1+\dfrac{r}{100}\right)^n$. (1)

It then follows that $n = \dfrac{\log 2}{\log\left(1+\dfrac{r}{100}\right)}$. (2)

Let us make a table of values (table 3.2) from the above equation with the help of a scientific calculator:

r	n	nr
1	69.66071689	69.66071689
3	23.44977225	70.34931675
5	14.20669908	71.03349541
7	10.24476835	71.71337846
9	8.043231727	72.38908554
11	6.641884618	73.0607308
13	5.671417169	73.72842319
15	4.959484455	74.39226682

Table 3.2.

If we take the arithmetic mean (the usual average) of the nr values, we get 72.04092314, which is quite close to 72, and so our "Rule of 72" seems to be a very close estimate for doubling money at an annual interest rate of r percent for n interest periods.

An ambitious reader or one with a very strong mathematics background might try to determine a "rule" for tripling and quadrupling money, similar to the way we dealt with the doubling of money. The above equation (2) for k-tupling would be

$$n = \dfrac{\log k}{\log\left(1+\dfrac{r}{100}\right)},$$

which for $r = 8$, gives the value for $n = 29.91884022$ (log k). Thus $nr = 239.3507218 \log k$, which for $k = 3$ (the tripling effect) gives us $nr = 114.1993167$. We could then say that for tripling money we would have a "Rule of 114."

Although we use computer calculation, for the most part, to determine compound interest, here we see a simple trick that allows us to skip a sizable portion of calculation and still come up with a useful answer. Once again, we see how mathematics helps us financially.

PAPER SIZES AND THE ROOT OF ALL ISO

Hardly a day goes by when we don't come into contact with the typical sheet of paper. Although this paper seems common, there are some fascinating mathematical relationships that can be noticed in the dimensions of the paper itself. Let's consider some of those now. The most common paper sizes in the United States are the "letter," the "legal," and the "ledger," also known as "tabloid." Their exact dimensions, as shown in table 3.3, were defined by the American National Standards Institute (ANSI).

	United States Paper Sizes				
	Length		**Width**		**Aspect Ratio (Rounded)**
Letter	11 in	279.4 mm	8.5 in	215.9 mm	1.294
Legal	14 in	355.6 mm	8.5 in	215.9 mm	1.647
Ledger/Tabloid	17 in	431.8 mm	11 in	279.4 mm	1.545

Table 3.3.

The size of a "ledger"/"tabloid" is exactly twice the size of a "letter," but their aspect ratios (i.e., the ratio of the width to the length) are different. While the ANSI paper sizes are also standard in Canada, all other countries in the world have adopted a different system of standard sizes defined by the International Organization for Standardization (ISO). In the ISO system, there are three series of paper sizes, named "A," "B," and "C." The sizes specified in each series have a very convenient property that the ANSI paper sizes do not have. First, by dividing an ISO sheet into two equal halves parallel to its shorter edges, we will obtain a sheet whose dimensions are exactly those of the next smaller size in this series, just as half a "ledger/tabloid" is the size of a "letter." But, unlike the ANSI paper sizes, the aspect ratio is the same for all formats in the ISO system. This is a particularly useful characteristic, since it means that sheets can be scaled up or down to other sizes without any margins or cutoffs. However, the scaling property of the ISO paper sizes requires a very special aspect ratio, and with just a little bit of mathematics we can easily find out what this special ratio must be.

Consider a rectangular sheet of paper with longer side a and shorter side b, as shown in figure 3.1. Thus, the aspect ratio is $\frac{a}{b}$. Now fold the sheet in half, parallel to its shorter edges. Clearly, the longer edge of the halves will be b and the shorter edge will be $\frac{a}{2}$.

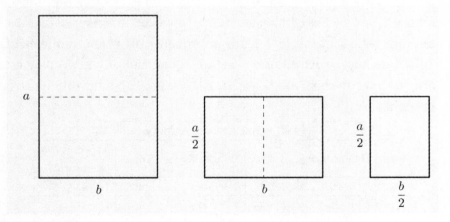

Figure 3.1.

108   THE MATHEMATICS OF EVERYDAY LIFE

Hence, the aspect ratio of the folded sheet is

$$b : \frac{a}{2} = \frac{b}{\frac{a}{2}} = 2\frac{b}{a}.$$

If we want this to be equal to the aspect ratio of the original sheet, we must have $\frac{a}{b} = 2\frac{b}{a}$. Multiplying this equation by $\frac{a}{b}$, we get $\left(\frac{a}{b}\right)^2 = 2$. This implies that $\frac{a}{b} = \sqrt{2}$ (where both a and b are positive). Therefore, only a sheet whose length-to-width ratio is exactly $\sqrt{2}$ can be folded or cut in halves (parallel to its shorter edge) without compromising this ratio.

The useful scaling property of paper sizes that are based on $\sqrt{2}$ was first recognized and advocated by the German physicist and philosopher Georg Christoph Lichtenberg (1742–1799). His idea was soon adopted in France, during the French Revolution. In the early twentieth century, a complete system of paper sizes was introduced in Germany that later turned into the ISO standard sizes. The dimensions of the five largest formats in the "A" series of the ISO paper sizes are shown in table 3.4. Theoretically, an A0 sheet is defined as a sheet with aspect ratio $\sqrt{2}$ and an area of 1 square meter. However, for practical reasons, the dimensions in the ISO norm are rounded to whole millimeters. Therefore, the aspect ratios are not all equal, and of course, they are also not exactly equal to $\sqrt{2}$ = 1.414213562 . . . , but the difference is insignificant.

ISO Series A Standard Sizes					
Size	Height (Longer Edge)		Width (Shorter Edge)		Aspect Ratio (Rounded)
A4	297 mm	11.7 in	210 mm	8.27 in	1.414286
A3	420 mm	16.5 in	297 mm	11.7 in	1.414141
A2	594 mm	23.4 in	420 mm	16.5 in	1.414286
A1	841 mm	33.1 in	594 mm	23.4 in	1.415825
A0	1189 mm	46.9 in	841 mm	33.1 in	1.413793

Table 3.4.

To illustrate an advantage of the scaling property of ISO paper sizes, let us consider a document printed on A4 sheets. Using a photocopier, we can reduce it to A5 so that two pages will fit on one A4 sheet. Since the aspect ratio stays the same, there will be no excess empty space. Analogously, we may magnify it to A3 without any waste of paper.

United States Paper Sizes					
Size	Long Edge		Short Edge		Aspect Ratio (Rounded)
ANSI A ("letter")	11 in	279.4 mm	8.5 in	215.9 mm	1.294118
ANSI B ("ledger"/"tabloid")	17 in	431.8 mm	11 in	279.4 mm	1.545455
ANSI C	22 in	558.8 mm	17 in	431.8 mm	1.294118
ANSI D	34 in	863.6 mm	22 in	558.8 mm	1.545455
ANSI E	44 in	1118 mm	34 in	863.6 mm	1.294118

Table 3.5.

In 1995, the American National Standards Institute adopted a system of paper sizes based upon the "letter" size and similar to the ISO system. Here, the "letter" size is represented by "ANSI A" and the "ledger"/"tabloid" by "ANSI B." Cutting a sheet in half produces two sheets of the next smaller size, but the aspect ratio is not preserved, since it was not equal to $\sqrt{2}$ in the first place. However, folding a sheet with sides a and b twice, we obtain a sheet with sides $\frac{a}{2}$ and $\frac{b}{2}$ (refer back to figure 3.1), whose aspect ratio is again $\frac{a}{b}$. So at least the aspect ratio is the same for every second paper size in this series.

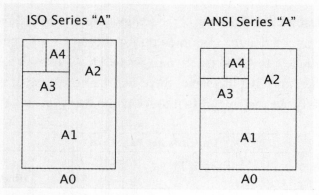

Figure 3.2.

The ISO paper sizes are a nice example of how elementary mathematical relationships can be used to produce an everyday item with a very convenient additional property that simplifies many standard procedures of printing or copying, thereby increasing their efficiency and eventually resulting in a cost reduction.

COMPARING AREAS AND PERIMETERS

Having now considered the dimensions of paper, let's take this one step further in a more general sense. As consumers, we are often required to make decisions about area and perimeter. This can come up in a number of ways. For example, if one is planning to construct a swimming pool and wants to maximize the surface area of the swimming pool, does that mean that the perimeters should be maximized? Furthermore, when installing drain pipes for the swimming pool, is one better off having one large drain hole with, say, a 6-inch diameter, or two drain holes side by side, each with 3-inch diameters? Some say in the latter case it would be the same. Some people might assume that the longer perimeter of a rectangle implies a larger area, and there is often the belief that the area of a rectangle would remain constant if the perimeter remains constant.

MATHEMATICAL APPEARANCES AND APPLICATIONS

An interesting situation was recently experienced when a contractor was planning the installation of ductwork in a new home and had to turn a right-angled corner with a duct whose cross-section was a square with a side length of 6 inches. When he told the owner that he was going to maintain the same perimeter but change the dimensions of the ductwork in order to make the right-angle turn so that the cross-section would be $3'' \cdot 9''$, the owner objected, fearing that it would restrict the airflow. Here was a situation where the contractor felt that as long as he kept the perimeter of the ductwork unchanged, the airflow would also be unchanged. Who was correct in this situation? As you will see during the following discussion, there may be some enlightening surprises ahead.

Comparing areas and perimeters is a very tricky thing. A given perimeter can yield many different areas. And a given area can be encompassed by many different perimeters. For example, let's consider rectangles of perimeter 20. As seen in figure 3.3, they may have very different areas despite their similar perimeters.

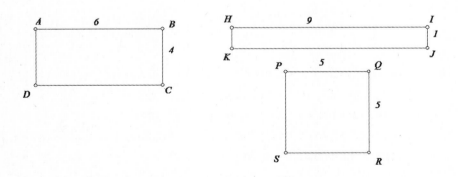

Figure 3.3.

With a perimeter of 20, the area of rectangle *ABCD* is 24.
With a perimeter of 20, the area of rectangle *HIJK* is 9.
With a perimeter of 20, the area of rectangle *PQRS* is 25.

It can be shown that the *maximum* area of a rectangle with a fixed perimeter is the one with equal length and width, that is, a square. Just the opposite is true for a rectangle of given area; the *minimum* perimeter is that where the length and width are equal, once again a square.

It is interesting to compare areas of similar figures. Next, we will consider circles. Suppose you have four equal pieces of string. They are shown in figure 3.4.

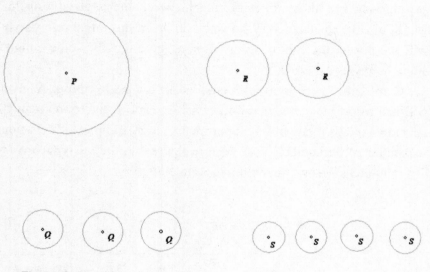

Figure 3.4.

With the first piece of string, one circle is formed.

The second piece of string is cut into two equal parts, and two congruent circles are formed.

The third piece of string is cut into three equal pieces, and three congruent circles are formed.

In a similar way, four congruent circles are formed from the fourth piece of string.

Note that the sum of the circumferences of each group of congruent circles is the same.

Circle	Diameter	Each Circle's Circumference	Sum of the Circles' Circumferences	Each Circle's Area	Sum of the Circles' Areas	Percent of Area of Circle P Represented by the Sum of the Areas of Smaller Circles
P	12	12π	12π	36π	36π	100
R	6	6π	12π	9π	18π	50
Q	4	4π	12π	4π	12π	$33\frac{1}{3}$
S	3	3π	12π	2.25π	9π	25

Table 3.6.

An inspection of table 3.6 indicates that the sum of the circumferences for each group of circles is the same yet the sum of the areas is quite different. The more circles we formed with the same total length of string, the smaller the total area of the circles. This is just what you would *not* intuitively expect to happen!

That is, when two equal circles were formed, the total area of the two circles was one-half that of the large circle. Similarly, when four equal circles were formed, the total area of the four circles was one-fourth of the area of the large circle.

From this counterintuitive perspective, if we consider a more extreme case, with, say, 100 smaller equal circles, we would see that the area of each circle becomes extremely small and the *sum* of the areas of these 100 circles is one-hundredth of the area of the larger circle. Try to explain this rather-disconcerting concept. It ought to give you an interesting perspective on the comparison of areas and perimeters.

MATHEMATICS IN HOME CONSTRUCTION

Needless to say, there are countless applications of mathematics in the home. Clearly, we cannot cover all of them, but we will select a few that give you a sense of mathematics at work. Let's begin with a problem that could come up in a number of applications. We will use one where a contractor comes into your home and wants to install some new ductwork. That is, to install a heating or air-conditioning passage through an aluminum channel (a duct) that has a rectangular cross-section. The situation described here actually confronted one of the authors. It demonstrates that the consideration of area can be extremely critical when involved with some home construction.

A contractor offered to replace some air-conditioning ducts at a friend's home. Needing to turn the duct around a corner at a right angle in the basement in a rather tight space, he indicated that he would keep the amount of aluminum the same, which was used to make a 6″ · 12″ duct, but as he would turn the corner, he claimed that he would need to make it narrower and use a 3″ · 15″ duct. He claimed that it should be no problem since the same amount of aluminum was being used. The author tried to explain to him that this was unacceptable because the cross-section area of the 3″ · 15″ duct would be considerably less than the 6″ · 12″ duct, which would have an effect of cutting down the airflow. The contractor didn't understand this point until the author chose to use as an example an extreme case, where the dimensions would be 0.5″ · 17.5″, which would clearly restrict the flow of air. After he was presented with this example, he realized the error of his ways.

What he had come to realize with the above example is that it is the *area* of the cross-section of the duct that determines the airflow. In other words, we had to compare the cross-section of a 6″ · 12″ duct with that of a 3″ · 15″ duct. In figure 3.5, we see the two cross-sections compared, where the first has an area of 6 · 12 = 72 square inches and the other has an area of 3 · 15 = 45 square inches. Clearly the latter allows less air to pass through than the former.

MATHEMATICAL APPEARANCES AND APPLICATIONS   115

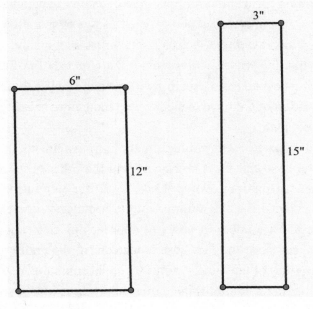

Figure 3.5.

For the given perimeter of this ductwork, 36″, the maximum area would be a square with a side length of 6″.

This can be extended to another situation, where we have a given amount of cardboard and we would like to create a box with the greatest possible volume from the given amount of cardboard. Rather than to experiment with different dimensions, we can extend the notion of a square providing the largest area from the given perimeter to see that a cube would generate the largest volume. That is, a rectangular solid with all square faces.

Another issue to deal with when it comes to home construction is the problem of emptying a swimming pool or a bathtub, where you have the option of using two drains of 2″ diameter or one drain of a 4″ diameter. Intuitively, it would seem that either option would drain equally quickly. However, that is not the case. Once again, we need to consider the cross-section of each of the two drain sizes. Using the famous formula for the area of the circle of radius 1, which is $\pi 1^2$, we find the area of the cross-section of each of the two smaller drains is equal to π, which together gives us a total cross-section area of 2π. On the other hand, the cross-section area of the larger drain of radius 2 is 4π, which is twice the drain capability of the two smaller drains. We see that our intuition might have misled us, but mathematics brings us back on the correct path.

An analogous problem would be that of filling a tub with two hoses. Suppose one hose can fill it in 2 hours and the other hose can fill the tub in 3 hours. How long would it take to fill the tub with both hoses working together? Once again, our intuition may not be very helpful here. There are a number of ways of approaching this problem. One is to notice that since the first hose alone can fill the entire tub in two hours, it can fill $\frac{1}{2}$ of the tub in 1 hour; while, in comparison, the second hose working alone could fill the entire tub in three hours, so it would fill $\frac{1}{3}$ of the tub in 1 hour. Working together, they would fill $\frac{1}{n}$th of the tub in one hour. Algebraically, we would represent that as follows: $\frac{1}{2}+\frac{1}{3}=\frac{1}{n}$, and then we have $n = \frac{6}{5}=1\frac{1}{5}$ hours, or one hour and 12 minutes. Again, this is something that is not intuitively obvious.

For this next illustration, the reader would be advised to briefly skip ahead to chapter 5 for the definition of a reflection in a plane so that we can use this concept to solve another home-construction problem. Suppose you would like to install an electric outlet along the wall, knowing that to this outlet two lamps will need to be connected. However, we would like to use the least amount of wire to connect these 2 lamps to the outlet. How might we locate this ideal point along the wall? Let's consider the situation shown in figure 3.6.

MATHEMATICAL APPEARANCES AND APPLICATIONS 117

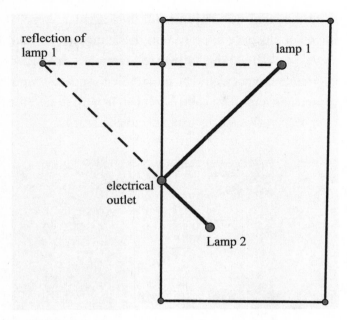

Figure 3.6.

One way of locating this ideal point for the electrical outlet is to find the reflection of lamp 1 in the wall, and then join that reflection point with a straight line to lamp 2. The point at which that line segment intersects the side of the wall is the point at which the outlet should be placed, since that is the minimum distance point from lamp 1 to the electrical outlet to lamp 2.

These are just some of the many applications of mathematics beyond those that we typically consider and that are usually found in the school curriculum.

THE PERFECT MANHOLE COVER

Suppose we now take our view of construction-related thinking outside of the house. How many times do we walk over manhole covers without

118   THE MATHEMATICS OF EVERYDAY LIFE

thinking for a moment why they are all in a circular shape? The vast majority of manhole covers we see on the street are, in fact, circles. Have you ever wondered why they are all circular? Analyze figure 3.7 to see why circular covers would be preferable to square-shaped covers. (As you can see, the circular shape cover can never fall into the hole, as would be the case with a square-shaped cover.)

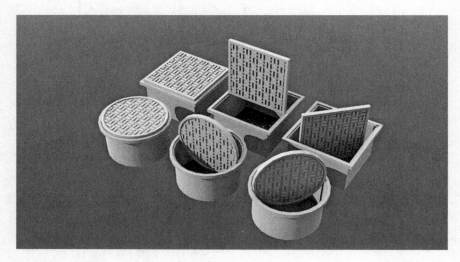

Figure 3.7.

The question then arises, is there any other shape that can be used to cover a manhole that will also not be able to fall into the hole? The answer to this question was provided by the German engineer Franz Reuleaux (1829–1905), who taught at the Royal Technical University of Berlin, Germany. He developed a rather odd-looking shape that is now called a *Reuleaux triangle*, which we show in figure 3.8.

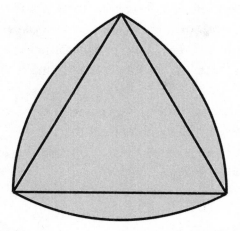

Figure 3.8.

One might wonder how Franz Reuleaux ever thought of this triangle. It is said that he was in search of a button that was not round but still could fit through a button hole equally well from any orientation. His "triangle" solved the problem. A manhole cover of the shape is shown in figure 3.9.

Figure 3.9.

120 THE MATHEMATICS OF EVERYDAY LIFE

The Reuleaux triangle is formed by three circular arcs along each side of an equilateral triangle with centers at the opposite vertices of the equilateral triangle. It has many unusual properties, and it compares nicely to a circle of similar breadth. In the case of a circle, the breadth is the diameter, while for the Reuleaux triangle it is the distance across—from a triangle vertex to the opposite arc. We refer to the distance between two parallel lines tangent to the curve as the breadth of the curve. In figure 3.10 we notice how a wrench is simply ineffectual when trying to turn a circular screw.

Figure 3.10.

The same would hold true for a Reuleaux triangular head (figure 3.11). It, too, would slip, since it is a curve of constant breadth, just as the circle is.

Figure 3.11.

This becomes particularly significant when fire hydrants are designed so that only special tools can turn them on. Often, a pentagonal valve screw is used to open such a hydrant, since a common wrench cannot be used, as would be the case with a hexagonal or a square valve screw, each of which has a pair of opposite parallel sides. Similarly, a circular valve screw would not suffice, as a wrench would not be able to grab it appropriately, as we saw in figure 3.10.

This brings us to the point of using a Reuleaux-triangle-type of valve screw, which can only be turned with a wrench of exactly the same shape. One such example of a fire hydrant is shown in figure 3.12.

Figure 3.12.

So here is a practical application of this situation. During the summer months, kids in a city like to "illegally" turn on the fire hydrants to cool off on very hot days. Since the valve of the hydrant is often a hexagonal-shaped nut, they simply get a wrench to open the hydrant. If that nut were the shape of a Reuleaux triangle, then the wrench would slip along the curve just as it would along a circle. However, with the

122 THE MATHEMATICS OF EVERYDAY LIFE

Reuleaux triangle nut, unlike a circular-shaped nut, we could have a special wrench with a congruent Reuleaux triangle shape that would fit about the nut and not slip. This would not be possible with a circular nut. Thus, the Fire Department would be equipped with a special Reuleaux wrench to open the hydrant in cases of fire, yet the Reuleaux triangle could protect against playful water opening, and avoid water being wasted in this manner.

A curious property of the Reuleaux triangle is that the ratio of its perimeter to its breadth, which is $\frac{\frac{1}{2} \cdot 2\pi r}{r} = \pi$, is the same as that ratio for a circle, $\frac{2\pi r}{2r} = \pi$.

The comparison of the areas of these two shapes is quite another thing, and it could be useful when deciding what shape to make a manhole cover. Let's compare their areas. We can get the area of the Reuleaux triangle in a clever way, by adding the three circle sectors that overlap in the equilateral triangle and then deducting the pieces that overlap, so that this region is actually only counted once and not three times.

The total area of the three overlapping circle sectors, where each is $\frac{1}{6}$ of the area of the circle shown in figure 3.13, is equal to $3\left(\frac{1}{6}\right)(\pi r^2)$. From this we need to subtract twice the area of the equilateral triangle, which is $\frac{r^2\sqrt{3}}{4}$. Therefore, the area of the Reuleaux triangle is equal to

$$3\left(\frac{1}{6}\right)(\pi r^2) - 2\left(\frac{r^2\sqrt{3}}{4}\right) = \frac{r^2}{2}(\pi - \sqrt{3}) \approx r^2\left(\frac{3.1416 - 1.732}{w}\right) = 0.7048r^2,$$

while the area of a circle with diameter of length r is equal to $\pi\left(\frac{r}{2}\right)^2 = \frac{\pi r^2}{4} = 07854r^2$.

Therefore, the area of the Reuleaux triangle is less than the area of the circle, which we can also see rather clearly in figure 3.13, where the two shapes are superimposed along with the an equilateral triangle.

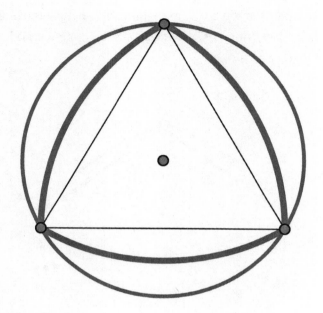

Figure 3.13.

This is consistent with our understanding of regular polygons, where the circle has the largest area for a given diameter. Furthermore, the Austrian mathematician Wilhelm Blaschke (1885–1962) proved that given any number of such figures of equal breadth, the Reuleaux triangle will always possess the smallest area, and the circle will have the greatest area.

Therefore, from a practical standpoint, to design a manhole cover of a given breadth, and one that would not be able to fall into the hole, the Reuleaux triangle shape would be the economic choice, as it would require less metal to construct. Here is an interesting application of some genuine mathematics applied to a piece of our environment that we seem to take for granted.

By the way, we can also create a Reuleaux pentagon by constructing circular arcs from each vertex and joining the endpoints of the opposite side, as shown in figure 3.14. This shape shares the properties of the

Reuleaux triangle and the circle in that it can fit between two fixed parallel lines regardless of its position. Can this be done with other regular polygons? If so, which ones?

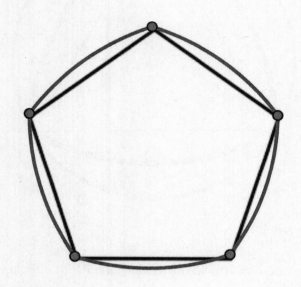

Figure 3.14.

DESIGN YOUR OWN COFFEE-CUP SLEEVE!

When you order a to-go coffee at a coffee shop or a fast-food restaurant, it will usually be served in a handle-less paper cup together with a tightly fitting cardboard sleeve, insulating your hand from the hot coffee, as we show in figure 3.15.

Figure 3.15.

The coffee-cup sleeve is a simple, yet ingenious, piece of design. It was patented in 1995 by Jay Sorensen (1958–), and it truly is a great example of industrial design. As "form follows function," it may not appear very impressive at first sight, but it had a huge effect on the coffee-to-go culture and soon became ubiquitous in modern coffee shops all around world. It was invented in response to a problem that almost everyone was concerned with: Hot coffee in a disposable, handle-less paper cup is not very enjoyable if your fingers get burned while drinking. Before the invention of the coffee-cup sleeve, using two or more nested paper cups for a single hot beverage was a common practice. Clearly, this is far from a reasonable solution, not only from an ecological viewpoint but also from an economical one. The coffee-cup sleeve represents a simple, cheap, and effective way to deal with this problem. Because the coffee-cup sleeve has become an indispensable item of everyday life, the Museum of Modern Art in New York acquired a standard coffee-cup sleeve for its collection "Humble

Masterpieces." An additional, often-neglected advantage of the coffee-cup sleeve is the possibility to reuse it, making it environmentally sustainable. However, manufactured coffee-cup sleeves are not very stylish. They often carry logos, brands, or even advertisements, and brown can be considered a rather-boring color. Why not design your own, reusable coffee-cup sleeve? Apart from protecting your fingers from the heat, this will also protect the environment, avoiding another disposable paper-cup sleeve every time you consume a "to-go coffee."

To create your personal coffee-cup sleeve, you need flexible cardboard or felt and a template as a guide to cut the material. The template can be printed or drawn on paper or cardboard and then cut out. By now, you may have started to wonder what this has to do with mathematics, but you will soon see. A template for a coffee-cup sleeve will look similar to the one we show in figure 3.16.

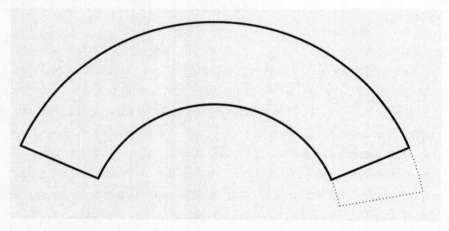

Figure 3.16.

It is basically a segment of a circular ring with an additional small segment that can be used as a fold to glue the ends together. If felt is used, the fold can be omitted since the ends would be sewed or pinned. But the important question we first have to answer is how to determine the proper shape of the template, given the shape of the coffee cup.

This is, in fact, a mathematical problem; more precisely, it is a problem belonging to elementary geometry. Of course, we could use brute force and cut open the cup to create a template out of it, but this wouldn't be very elegant. A paper cup is essentially a frustum of a cone, that is, the remainder of a cone when the vertex is cut off parallel to the base, determined by its diameters at the bottom and the top, and its height. In figure 3.17, we show a frustum of a cone with its profile, where the segment AB represents the top radius, r_1, and the segment $A'B'$ represents the bottom radius, r_2. We have also labeled the vertical height, $h = AA'$, and the lateral height, $s = BB'$, measured flat to the cup wall. To construct the apex of the cone, we extend the segments AA' and BB' to their point of intersection, C, and denote the segment $B'C$ by t.

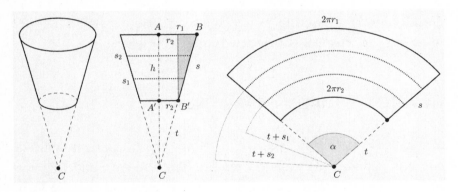

Figure 3.17.

Knowing t, we can construct the full lateral surface area of the frustum, which we also show in figure 3.17. It is a segment of a circular ring with inner radius t and outer radius $t + s$. The outer circular arc, at distance $t + s$ from the center, is of length $2\pi r_1$. The inner circular arc, at distance t from the center, is of length $2\pi r_2$. To determine the angle α, we only have to find the fraction of a full circle that is covered by these circular arcs. The circumference of a circle with radius t is $2\pi t$, hence the inner circular arc of length $2\pi r_2$ is a fraction, $\frac{r_2}{t}$, of the full circle and thus $\alpha = 360° \cdot \frac{r_2}{t}$.

128 THE MATHEMATICS OF EVERYDAY LIFE

To find t, we observe that the shaded triangle in the center of figure 3.17 is similar to triangle $A'B'C$. Therefore, we have $\frac{t}{r_2} = \frac{s}{r_1 - r_2}$, implying that $t = s \frac{r_2}{r_1 - r_2}$. Substituting t in the formula for α, we get $\alpha = 360° \cdot \frac{r_1 - r_2}{s}$.

Assume that the band of the sleeve shall be between s_1 and s_2, measured from the bottom of the cup along its lateral height. Then the template for this sleeve would be a segment of a circular ring with inner radius $t + s_1$ and outer radius $t + s_2$ (see fig. 3.17). The distance t and the angle α are all we need to draw the template.

In summary, to create a template for a tightly fitting coffee-cup sleeve, you have to measure the diameters on the top and the bottom of the cup and its lateral height. With this information, you can easily compute the required values of $t = s \frac{r_2}{r_1 - r_2}$ and $\alpha = 360° \cdot \frac{r_1 - r_2}{s}$. Then you choose the breadth of the band by specifying s_1 and s_2. This is all you need to draw the template. We hope you will enjoy your next coffee to-go in your personally constructed coffee-cup sleeve!

HOW TO OPTIMALLY WRAP A PRESENT

You're back home now, and you're faced with a common domestic concern. Suppose you want to wrap a present and are wondering how much wrapping paper you will need. If the shape of the item to be wrapped is that of a polyhedron (a three-dimensional shape that has flat surfaces and straight edges), then the minimum amount of paper you will need is represented by the total surface area of the polyhedron. Now imagine you cut the surface of the polyhedron along its edges, such that you can unfold it to get a non-overlapping arrangement of edge-joined polygons in the plane. This is called the "net" of the polyhedron. In figure 3.18, we show the net of a regular tetrahedron, a polyhedron composed of four equilateral triangles.

MATHEMATICAL APPEARANCES AND APPLICATIONS 129

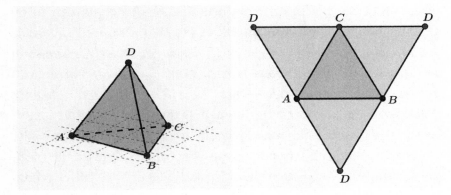

Figure 3.18.

Depending on the choices of which edges are joined and which are separated, you may obtain different nets for the same polyhedron. For instance, there exist eleven distinct nets for a cube, which we show in figure 3.19.

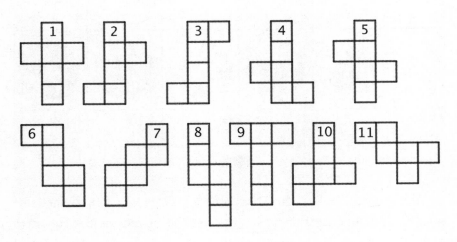

Figure 3.19.

This might be helpful if you want to wrap a cube and only have a small and irregularly shaped piece of wrapping paper left. Although it may seem impossible to wrap this leftover piece of paper all around the

130 THE MATHEMATICS OF EVERYDAY LIFE

cube, one of the eleven different nets of the cube could still fit on it. The German artist Albrecht Dürer (1471–1528) used such nets to show polyhedrons in his book *Underweysung der Messung mit dem Zirckel und Richtscheyt* (*Instructions for Measurement with Compass and Ruler*, also known as *Four Books on Measurement*). Determining the number of different nets that exist for a given polyhedron is by itself an interesting geometrical problem. It is actually not known whether every convex polyhedron has at least one (nonoverlapping) net. Although there is no evidence that Dürer recognized this statement as something that has to be proved, it is now called Dürer's conjecture. In figure 3.20 we show the number of distinct nets for the five Platonic solids (convex polyhedrons with congruent faces).

	Tetrahedron	**Cube (Hexahedron)**	**Octahedron**	**Dodecahedron**	**Icosahedron**
Distinct Nets	2	11	11	43380	43380

Figures 3.20.

For nonregular polyhedrons (not all faces are congruent), the number of distinct nets increases. For example, a rectangular cuboid (a solid comprised of three opposite pairs of congruent rectangular faces) with three different lengths a, b, c has fifty-four different nets. Therefore, finding the optimal net to make the best use of a leftover piece of wrapping paper is not an easy task. However, we will consider the more common situation that we have a roll of wrapping paper at hand

MATHEMATICAL APPEARANCES AND APPLICATIONS

and want to cut off a sheet of paper that is just large enough to wrap the item. Our goal is to avoid unnecessary waste, but, since we don't want to spend too much time for wrapping the present, we want to make only one straight cut to separate a rectangular piece from the roll. We will further assume that the item has the shape of a rectangular cuboid with sides $a \geq b \geq c$. Of the up to fifty-four possible nets, we are interested in variants closely resembling a rectangle, in order to minimize excess paper. Two nets of the same rectangular cuboid with sides $a \geq b \geq c$ are shown in figure 3.21.

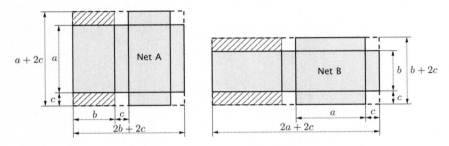

Figure 3.21.

If we ignore the width of the paper roll, the minimum size of the wrapping paper is given by the smallest rectangle that contains a net of this cuboid. Net A fits inside a rectangle with sides $2b + 2c$ and $a + 2c$. (See fig. 3.21.) The area of this rectangle is therefore $2(b + c)(a + 2c)$. The area of the smallest rectangle containing net B is $2(a + c)(b + 2c)$. (See fig. 3.21.) It is not immediately obvious which area is smaller (we would have to expand and subtract the two expressions). However, instead of comparing the areas of the enclosing rectangles, we may also look at the area of the excess paper (since the areas of the nets must be equal). As we can see in figure 3.21, the excess area for each of the nets can be decomposed into four identical squares of size c^2 and two rectangles of height c, but with different lengths. The rectangles alongside net A are b units long, while those alongside net B are a units long.

Since $b \leq a$, there is less waste if we cut out a rectangular piece of paper enclosing net A. The total excess area of $2c(b + 2c)$ is actually not completely wasted, since we may need some overlap to affix the wrapping paper appropriately anyway.

So far we have not included the width of the wrapping paper roll in our considerations. This will change the picture, since the length of one side of the rectangular piece of paper is already fixed by the width w of the roll. Both nets can be placed either parallel ("horizontal") or perpendicular ("vertical") to the longer side of the rolled-up paper. We show all four possible arrangements in figure 3.22.

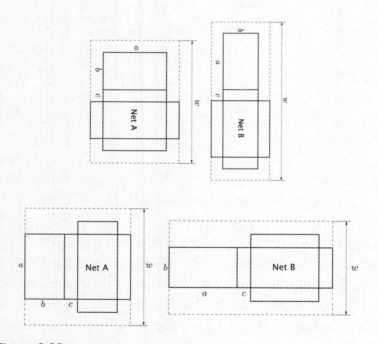

Figure 3.22.

As one side of the dashed-line rectangle has always length w, the best choice is the arrangement for which the other side of the rectangle is the shortest, and the answer depends on the dimensions of the cuboid in relation to the width of the roll. If net B can be placed "vertically," which is only possible for $w \geq 2(a + c)$, then this would be the most efficient method (see figure 3.22). If $w < 2(a + c)$, and $2b > a$, you should first try to place net A vertically or, if this is not possible, horizontally. If $2b < a$, then you should try the horizontal orientation first. The final option is to place net B horizontally. Assuming that c is much smaller than w and therefore $w - 2c \approx w$ (which would be the case if the item to be wrapped is a book, for instance), in table 3.7 we show the best choice and orientation for different ranges of w.

$2b > a$	$w > 2a$	$2a > w > 2b$	$2b > w > a$	$a > w > b$
	B vertical	A vertical	A horizontal	B horizontal
$a > 2b$:	$w > 2a$	$2a > w > a$	$a > w > 2b$	$2b > w > b$
	B vertical	A horizontal	A vertical	B horizontal

Table 3.7.

If $a = b$ and c is much smaller than a, that is, if the cuboid has the shape of a relatively flat square, then there exists an alternative method to wrap the item. This method not only is more efficient but also makes the present look nicer. You need a square piece of paper with sides of length $\sqrt{2}\,(a + c)$, where $\sqrt{2} = 1.414...$, which we can round up to 1.42 here (or, being a bit more generous, to 1.45). You then fold the paper in half two times and in alternate directions to get guidelines for placing the item (see figure 3.23). Now the trick is to place the cuboid in the middle of the square piece of paper with its corners on the guidelines, as we show in figure 3.23, and fold up the corners of the wrapping paper. They all should meet in the center, and you need only a tiny bit of tape to affix them (the small amount of excess paper along the sides of length c is easy to cover). There are no overlapping folds of wrapping paper needed, which makes the present look particularly nice.

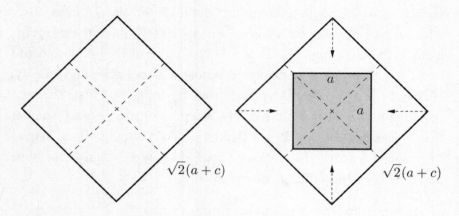

Figure 3.23.

At this point, we hope that you will have become a much more clever consumer, both geometrically and arithmetically. However, cleverness has another role in our mathematical world, and that involves probability. Continue with this on our journey as we enter another realm of mathematics that helps us navigate our world by making the right decisions in games, gambling, and beyond.

CHAPTER 4

PROBABILITY, GAMES, AND GAMBLING

The concept of probability plays a very important role in our society today. Yet it all began in 1654 when two famous French mathematicians, Blaise Pascal (1623–1662) and Pierre de Fermat (1601–1665) corresponded about the chances of winning in a then-popular dice game.[1] The game consisted of tossing a pair of dice 24 times to determine whether to bet even money that at least one of those 24 tosses will result in a pair of sixes. It was believed then that getting a pair of sixes during these 24 throws was a good bet, yet the opposite resulted from the correspondence between these two famous mathematicians. They further corresponded about such gaming outcomes, which eventually led to determining the mathematical field of probability, which was formalized in 1657 by the first book published in this field by the Dutch mathematician Christian Huygens (1629–1695), titled *De Ratiociniis in Ludo Aleae*. The continued study of probability has led to many unusual results, which we shall consider in this chapter.

FRIDAY THE THIRTEENTH!

The number 13 is usually associated with being an unlucky number. Buildings with more than thirteen stories typically will omit the number 13 from the floor numbering. This is immediately noticeable in the elevator, where there is no button for 13. You can certainly think of other examples where the number 13 is associated with bad luck. This fear of the number 13 is often referred to as *triskaidekaphobia*. Among the more famous people in history who suffered from triskaidekaphobia are Franklin D. Roosevelt, Herbert Hoover, and Napoleon Bonaparte.[2]

You will likely recall that when the 13th of a month turns up on a Friday, there are some folks who believe that there could be particularly bad luck on that day. This may derive from the belief that there were thirteen people present at the Last Supper, which resulted in Jesus's crucifixion on a Friday. Do you think that the 13th falls on a Friday with equal regularity as it does on the other days of the week? You may be astonished that, lo and behold, the 13th comes up more frequently on Friday than on any other day of the week.

This fact was first published by the American mathematician Bancroft H. Brown (1894–1974) in *American Mathematical Monthly*.[3] He stated that the Gregorian calendar follows a pattern of leap years, repeating every 400 years. The number of days in one four-year cycle is $3 \cdot 365 + 366$. So, in 400, years there are $100(3 \cdot 365 + 366) - 3 = 146,097$ days. Note that the century year, unless divisible by 400, is not a leap year; hence the deduction of 3. This total number of days is exactly divisible by 7, which you can determine by using the divisibility rule shown in chapter 1. Since there are 4,800 months in this 400-year cycle, the date of the 13th comes up 4,800 times. Interestingly enough, the 13th comes up on a Friday more often than on any other day of the week. The following chart (table 4.1) summarizes the frequency of the 13th appearing on the various days of the week. Here we have a conundrum: the day that occurs most frequently is considered bad luck!

Day of the Week	Number of 13s	Percent
Sunday	687	14.313
Monday	685	14.271
Tuesday	685	14.271
Wednesday	687	14.313
Thursday	684	14.250
Friday	*688*	*14.333*
Saturday	684	14.250

Table 4.1.

There are some people who were not consciously affected by the number 13 but have had much of their lives involving this rather-famous number. For example, the famous German opera composer Richard Wagner was born in 1813 (and the sum of the digits of this year is 13), and his name consists of 13 letters.[4] Wagner was first motivated toward his life's work at a performance of Carl Maria von Weber's opera *Der Freischütz* on October 13, 1822. Wagner composed 13 operas. One of his operas, *Tannhäuser*, was completed on April 13, 1845, and was first performed in Paris during his exile from Germany on March 13, 1861. He spent 13 years in exile for political reasons. Wagner's last day in the city of Bayreuth, Germany, where he had built his famous opera house, was September 13, 1882. His father-in-law, the Hungarian composer Franz Liszt (1811–1886), saw Wagner for the last time on January 13, 1883. Wagner died on February 13, 1883, which just happens to be the 13th year of the unification of Germany. Curiosities such as these bring another dimension of entertainment to numbers.

UNEXPECTED BIRTHDAY MATCHES

In our everyday lives, we might consider the likelihood of something happening or not happening. This often goes under the heading of how probable or how likely something is to happen. The topic of probability can explain many things. Yet one of the most surprising results in mathematics—and one that is quite counterintuitive—is the question of how likely it is for two people to have the same birthdate (just month and day, ignoring the year). When you find out the likelihood of this happening, it will surely upset your sense of intuition. On the other hand, it is one of the best ways to convince the uninitiated about the "power" of probability.

Let us suppose that you are in a group with 35 other people. What do you think are the chances (or probability) of two people in the group having the same birthdate (month and day, only)? Intuitively, one usually begins to think about the likelihood of 2 people having the same date out of a selection of 365 days (assuming no leap year). Perhaps 2 out of 365? That would be a probability of $\frac{2}{365} = .005479 \approx \frac{1}{2}\%$. A rather-minuscule chance.

Rather than our group of 35 people, let's consider the "randomly" selected group of the first 35 presidents of the United States. You may be astonished that there are two with the same birthdate: the eleventh president, James K. Polk (November 2, 1795), and the twenty-ninth president, Warren G. Harding (November 2, 1865).

You may be surprised to learn that for a group of 35 people, the probability that two of them will have the same birthdate is greater than 8 out of 10, or $\frac{8}{10} = 80\%$.

If you have the opportunity, you may wish to try your own experiment by selecting 10 groups of about 35 members in each group and check for birthdate matches in each group. For groups of 30 people, the probability that there will be a match of birthdates is greater than 7 out of 10, or, put another way, in 7 of these 10 groups there ought to be a match of birthdates. What causes this incredible and unanticipated result? Can this really be true? It seems to go against our intuition.

To relieve you of your curiosity, we will consider the situation in detail.

Let's consider a group of 35 people. What do you think is the probability that one selected person matches his/her own birthday? Clearly *certainty*, which we express as a probability of 1. This can be written as $\frac{365}{365}$.

The probability that another person in the group does *not* match the first person is $\frac{365-1}{365} = \frac{364}{365}$.

The probability that a third person does *not* match the first and second person is $\frac{365-2}{365} = \frac{363}{365}$.

The probability of all 35 people *not* having the same birth date is the product of these probabilities:

$$p = \frac{365}{365} \cdot \frac{365-1}{365} \cdot \frac{365-2}{365} \cdot \ldots \cdot \frac{365-34}{365}.$$

Since the probability (q) that two people in the group *do have* the same birthdate and the probability (p) that two people in the group *do not have* the same birthdate is a certainty, the sum of those probabilities must be 1. Thus, $p + q = 1$.

In this case,

$$q = 1 - \frac{365}{365} \cdot \frac{365-1}{365} \cdot \frac{365-2}{365} \cdot \ldots \cdot \frac{365-33}{365} \cdot \frac{365-34}{365} \approx 0.8143832388747152.$$

In other words, the probability that there will be a birthdate match in a randomly selected group of 35 people is somewhat greater than $\frac{8}{10}$. This is quite unexpected when you consider there were 365 dates from which to choose. If you are feeling motivated, you may want to investigate the nature of the probability function. Table 4.2 may further enlighten you.

Number of People in Group	Probability of a Birthdate Match
10	0.1169481777110776
15	0.2529013197636863
20	0.4114383835805799
25	0.5686997039694639
30	0.7063162427192686
35	0.8143832388747152
40	0.891231809817949
45	0.9409758994657749
50	0.9703735795779884
55	0.9862622888164461
60	0.994122660865348
65	0.9976831073124921
70	0.9991595759651571

Table 4.2.

Notice how quickly "almost certainty" is reached. With about 60 people in a room, the chart indicates that it is almost certain (0.99) that two of them will have the same birthdate.

Were you to do this with the death dates of the first 35 presidents, you would notice that two died on March 8 (Millard Fillmore in 1874 and William H. Taft in 1930), and three presidents died on July 4 (John Adams and Thomas Jefferson in 1826, and James Monroe in 1831). How curious that the only two signers of the Declaration of Independence who went on to become presidents (Adams and Jefferson) died on the very same day—the date that the Declaration the Independence was ratified!

Above all, this astonishing demonstration should serve as an eye-opener about the inadvisability of relying too much on intuition, and becoming aware of how mathematics is truly a part of our lives.

SELECTING CLOTHES

There are times when selecting clothes can be seen as a mathematical problem to be solved. One might also consider this an exercise in logical thinking. Here we see an example that shows how reasoning with extremes, as we discussed in an earlier chapter, is a particularly useful strategy to solve some problems. It can also be seen as a "worst-case scenario" strategy. The best way to appreciate this kind of thinking is through example. Therefore, we will consider the following problem.

> *The problem:* **In a drawer, there are 8 blue socks, 6 green socks, and 12 black socks. What is the fewest number of socks that Charlie must take from the drawer in a dark room—where he cannot see the colors—to be certain that he has selected two socks of the same color?**

The problem does not specify which color, so any of the three colors would satisfy his selection. To solve this problem, you might reason from the perspective of a "worst-case scenario." Suppose Charlie is very unlucky and on his first 3 picks, he selects one blue sock, one green sock, and then one black sock. He now has one of each color, but no matching pair. Although he might have picked a matching pair on his first two selections, we need to determine how many socks he must select to be *certain* of having a matching pair. As soon as he now picks the fourth sock, he must match one of his first three picks; and, therefore, he would have a pair of the same color.

Let's go back to the sock drawer and consider an alternative problem situation as expressed in the following problem:

> *The problem:* **In a drawer, there are 8 blue socks, 6 green socks, and 12 black socks. What is the fewest number of socks that Max must take from the drawer—without looking—to be certain that he has selected two black socks?**

Although this problem appears to be similar to the previous one, there is one important difference. In this problem, a specific color has been required: Max must choose a pair of black socks. Again, let's use deductive reasoning and construct the "worst-case scenario." Suppose Max first picks all of the eight blue socks. Next, he picks all six green socks. Still not one black sock has been chosen. Max now has selected 14 socks in all, but none of them is black. However, the next two socks he picks must be black, since there are only black socks remaining in the drawer. In order to be certain of picking two black socks, Max must select $8 + 6 + 2 = 16$ socks in all.

There are other considerations when selecting clothes to wear, and that requires what is known in mathematics as the fundamental accounting principle. Suppose we are traveling and have taken along 5 shirts, 3 pairs of pants, and 2 jackets. The question then is, How many different outfits can we make from the clothes we have taken along? The reasoning is that for each of the 3 pairs of pants we can use any one of 5 shirts, which means there are 15 possible arrangements of these garments. Then for each of the 15 possible arrangements we can select one of the 2 jackets, providing us with 30 possible outfits. What the fundamental accounting principle tells us is that we merely have to multiply the three numbers to get the total number of outfits.

PLAYING CARDS, A COUNTERINTUITIVE PROBABILITY

The topic of probability permeates all of card playing. A knowledge of the probability of getting a winning hand at any card game will always give an advantage to the person betting on the game. There are entire books written about the probability of getting certain poker hands or winning at blackjack; however, we will take a very simple situation with two decks of cards (each deck containing 52 cards) to show how a simple question can lead to a very significant mathematical constant. In addition, the result we show will probably be counterintuitive.

PROBABILITY, GAMES, AND GAMBLING

The problem: **If you have two well-shuffled decks stacked face-down beside each other and begin by turning over one card from each deck at the same time, what is the probability that at least one pair being turned over will be identical?**

We begin with the two decks of cards, but to make our process simpler, we will shuffle only one deck of cards, and the other deck will be placed in a sequential numerical order. The probability that the first card turned over on each deck will be a match is $\frac{1}{52}$, since there are 52 cards in the shuffled deck and only one of them will match the first card of the ordered deck. From this we can conclude that the probability that the two first cards *don't* match is $1 - \frac{1}{52}$. Moving along, the probability that each succeeding turned over card does not match its partner is approximately $1 - \frac{1}{52}$. We say *approximately* because we are ignoring the fact that an earlier card in the shuffled deck could have come up for its partner in the ordered deck. We thus expect the probability that none of the 52 cards turned over in pairs produced a match is the product of these 52 individual probabilities of approximately $1 - \frac{1}{52}$, which is actually $\left(1 - \frac{1}{52}\right)^{52}$.

This expression reminds us of a very significant expression in mathematics, which we encountered earlier. The general form would be $\left(1 + \frac{1}{n}\right)^n$. When this value of n takes on increasingly larger numbers, we find that it approaches $2.718281828459045235360287471352 7\ldots$, which is referred to by the famous Swiss mathematician Leonhard Euler (1707–1783) with the letter e, and it is one of the most important mathematical constants. Another way of generating the value of e is the following series: $1 + \frac{1}{1} + \frac{1}{1 \cdot 2} + \frac{1}{1 \cdot 2 \cdot 3} + \frac{1}{1 \cdot 2 \cdot 3 \cdot 4} + \ldots + \frac{1}{n!}$.

Let's now consider e^x, which will give us $\left(1 + \frac{x}{n}\right)^n$. Reverting to our original expression that generated our digression to e, namely, $\left(1 - \frac{1}{52}\right)^{52}$, we find that this is very close to $e^{-1} = \frac{1}{e}$. This is approximately $\frac{1}{2.718}$, a number somewhat greater than $\frac{1}{3}$, which is the probability that no identical pairs will evolve from the two decks of cards. Or, stated another way, the probability that an identical pair *will* evolve is somewhat less than two-thirds. This is truly for most people counterintuitive, and, further, it shows how probability becomes an integral part of higher mathematics, as it invoked the famous constant e. By the way, we should note that Euler was also responsible for labeling the ratio of the circumference to the diameter of a circle with the Greek letter π. Both numbers are irrational, since they cannot be expressed as the quotient of two integers. As with the value π, the value e is also considered a *transcendental number*, since it is not the root of any non-zero polynomial with rational coefficients.

MATHEMATICS IN POKER

Now that we have a basic grip on the concept of probability, let's apply it to perhaps one of the most popular card games, five-card poker. As a refresher, we shall quickly review the game in its most simple fashion. Each player is dealt five cards, and the one who holds the best hand wins. In descending order the best hands are: a royal flush (a straight up to the ace in one suit), four of a kind, a straight flush, full house, flush, straight, three-of-a-kind, two-pair, one-pair, and the highest card.

To begin, let us consider the number of possible poker hands that one could have been dealt from a deck of 52 cards. In other words, we seek to find the number of combinations of 5 cards possibly dealt from a deck of 52 cards. We represent that as $_{52}C_5$, which is $\frac{52 \cdot 51 \cdot 50 \cdot 49 \cdot 48}{1 \cdot 2 \cdot 3 \cdot 4 \cdot 5} = 2,598,960$.

Royal Flush

Let's now consider each of the above-listed winning poker hands in order according to their likelihood of occurrence, or, put another way, their probability of happening. The royal flush consists of the ace, king, queen, jack, and 10, all of the same suit. Suppose we consider the ace of spades. There is then only one possible group of four cards that would complete a royal flush in spades. However, there are 4 suits that qualify for a royal flush. Thus, there are only four possible royal flushes that one could get. The probability of a royal flush is, therefore, $P(\text{royal flush}) = \frac{4}{2,598,960} = 1.53907716932927 \cdot 10^{-6} = 0.00000153907716932927$, or about 0.000154 percent. This indicates odds of 1 in 649,739 chances, written as 1:649,739. To obtain these odds, we calculate in the following way: the four successful events as compared to the unsuccessful events $2,598,960 - 4 = 2,598,956$, which can be written as 4:2,598,956; then, dividing by 4, we get: 1:649,739.

Straight Flush

To find the probability of getting a straight flush (excluding the royal flush), which is having all five cards in consecutive order of the same suit, excluding the highest five consecutive numbers, we will order the cards from highest to lowest. We begin with the king, because if we use the aces as the highest, then the five consecutive cards would represent a royal flush, which we already considered above. There are then nine possibilities for a straight flush in each of the 4 suits, that is, $9 \cdot {}_4C_1 = 9 \cdot 4 = 36$. Therefore, the probability of getting a straight flush is $P(\text{straight-flush}) = \frac{36}{2,598,960} = 1.38516945239634 \cdot 10^{-5}$, or 1 chance in 72,192 tries.

Four of a Kind

We're now ready to find the probability of obtaining *four of a kind*, which is four cards showing the same number or letter. Suppose we arrange the cards in a five-card hand such that the four of a kind will be the first four cards. There are 13 possible such hands, and there are 48 cards remaining in the deck to sell the fifth position in the hand. Therefore, we have $13 \cdot 4 \cdot 48 = 624$ possible hands of four of a kind. Thus, the probability of obtaining four of a kind is $P(\text{four of a kind}) = \frac{624}{2,598,960} =$ .00024009603842, which yields odds of 1:4,164.

Full House

Now we want to find the probability of getting a full house, which consists of obtaining three cards of one number or letter and two cards of another. There are $_{13}C_1 = 13$ ways of getting the first number of the three-of a kind, and there are three out of four possible suits, which yields $_4C_3 = \frac{4 \cdot 3 \cdot 2}{1 \cdot 2 \cdot 3} = 4$ possibilities. For the pair that will complete this hand, we have $_{12}C_1 = 12$ possibilities of the number that can be shown on the two cards. As for the suits of this pair, we have $_4C_2 = 6$ possibilities.

We then have the probability of getting a full house as $P(\text{full house}) = {}_{13}C_1 \cdot {}_4C_3 \cdot {}_{12}C_1 \cdot {}_4C_2 = 13 \ 4 \ 12 \ 6 = 3744$. To determine the probability of getting a full house, we divide that number by the total number of possible hands: $\frac{3744}{2,598,960} = .0001440576$, or, expressed as odds, 1 : 693.

Flush

Now we move along to our next more-likely hand to obtain, which is having five cards of the same suit. This is called a flush, excluding the straight flush that we already considered above. There are 4 choices to pick a suit. Within that suit of 13 cards, the number of ways we can choose five cards is $_{13}C_5 = \frac{13 \cdot 12 \cdot 11 \cdot 10 \cdot 9}{1 \cdot 2 \cdot 3 \cdot 4 \cdot 5} = \frac{154,440}{120} = 1,287$. From this number, we need to subtract the 10 straights that might occur, so that we get the number of flushes as: $_4C_1 \cdot (_{13}C_5 - 10) = 4 \cdot 1277 = 5,108$. In order to calculate the probability of getting a flush, we divide this number by the total number of possible hands dealt, which is 2,598,960. Therefore, $P(\text{flush}) = \frac{5,108}{2,598,960} = .001965401$, and so the chances of getting a flush are about 1 in 508, or written as odds, 1:508.

Straight

As expected, we will now consider the probability of getting a straight, yet we'll exclude the straight flush. Once again, we see that there are 10 possibilities of getting a straight, where the ace can be either the highest card or the lowest card. Once we know what the first card is, the remaining cards are clearly determined. Since each of the cards can be of a different suit, each can be selected in four ways. Therefore, the number of straights obtainable is equal to $10 \cdot 4 \cdot 4 \cdot 4 \cdot 4 \cdot 4 = 10{,}240$. From this we need to subtract the number of straight flushes and royal flushes, which gives us $10{,}240 - 36 - 4 = 10{,}200$. Therefore, the probability of getting a straight (but not a flush) is $P(\text{non-flush straight}) = \frac{10{,}200}{2{,}598{,}960} = .003924646$, which translates to approximately one chance in 255 possibilities, or 1:254. The odds are defined as the ratio "(chances for):(chances against)"; therefore, one chance in 255 possibilities amounts to odds of 1:254, as there are $255 - 1 = 254$ chances against the event. We didn't subtract the 1 when we calculated the odds for the better poker hands we considered before, since the difference between both ratios would have been negligible. However, as we move on to more-likely poker hands, we can't ignore the difference between probability and odds anymore.

Three of a Kind

The next more-likely hand to obtain is to get three of a kind among the five cards dealt, with the remaining two cards not matching in number. Let's consider the first of the 3 cards that will be matching a number. There are 13 possible numbers that this first card could have. We will also have $_4C_3 = 4$ ways in which we can select the suits of these three cards. The remaining 2 cards in our hand must be different from the first 3 cards and different from each other, which implies that there are $_{12}C_2$ choices for these cards, in addition to which each of these cards can be of any one of 4 suits. Therefore, the number of possible hands containing three of a kind is $_{13}C_1 \cdot {}_4C_3 \cdot {}_{12}C_2 \cdot {}_4C_1 \cdot {}_4C_1 = 13 \cdot 4 \cdot 66 \cdot 4 \cdot 4 = 54{,}912$. When we divide this number by the total number of possible poker hands, we find the probability of getting three of a kind to be $P(\text{three of a kind}) = \frac{54{,}912}{2{,}598{,}960} = .021128451$, which translates to about 1 in 47 chances for getting a hand with three cards of the same number or letter, or odds of 1:46.

Two Pairs

As we move along, we find even more likely poker hands. This time we will consider the probability of getting a five-card hand dealt where there are 2 pairs of the same number or letter. To begin, there are $_{13}C_2$ ways that the 2 numbers can be selected to represent each of the 2 pairs. Within each pair, there can be 2 out of the 4 suits, or $_4C_2 \cdot {}_4C_2 = 6 \cdot 6 = 36$ ways to be selected. To avoid having a full house, the fifth card must be different from these other 2 pairs. There are $_{11}C_1$ ways to select a fifth card, and it can have any one of the 4 suits. To calculate the number of ways of getting two pairs, we calculate the following: $_{13}C_2 \cdot {}_4C_2 \cdot {}_4C_2 \cdot {}_{11}C_1 \cdot {}_4C_1 = 78 \cdot 6 \cdot 6 \cdot 11 \cdot 4 = 123{,}552$. Once again, in order to find the probability of getting a poker hand with 2 pairs, we divide this latter number by the total number of hands possible to get: $P(\text{two pairs}) = \frac{123{,}552}{2{,}598{,}960} = .047539015$, which yields odds of about 1 to 20.

One Pair

By now you may have noticed that the odds are continuously becoming more favorable, or, in other words, you are more likely to get these hands as we move along. Perhaps the simplest, or perhaps the most likely hand one can be dealt is getting one pair of cards, showing the same number or letter, with the other three cards in your hand different from the first two and from each other. Let's consider the pair of cards matching in number or letter. There are 13 possible ways that this can have the same number or letter ($_{13}C_1$). The pair will represent 2 of the 4 available suits, or $_4C_2 = 6$ ways that the suits can be selected. The rest of the cards in your hand must not match the first two cards and can be selected in $_{12}C_3$ ways, and may have any one of the 4 suits.

Therefore, the number of ways of getting exactly one pair of matching cards is as follows: $_{13}C_1 \cdot {}_4C_2 \cdot {}_{12}C_3 \cdot {}_4C_1 \cdot {}_4C_1 \cdot {}_4C_1 = 13 \cdot 6 \cdot 220 \cdot 4 \cdot 4 \cdot 4 = 1,098,240$. Once again, when we divide this number by the total number of possible hands, we get the probability of being dealt a pair of matching cards, which is $P(\text{one pair}) = \frac{1,098,240}{2,598,960} = .422569027$. This allows us to calculate that the odds of being dealt one pair is about 1:1.37.

Five Different Cards

The only other possible hand one can be dealt in poker is one in which there is none of the above occurrences. Therefore, there must be 5 different cards that can be selected in $_{13}C_5$ ways, and each of the cards can have any one of the 4 suits, yielding $_{13}C_5 \cdot {}_4C_1 \cdot {}_4C_1 \cdot {}_4C_1 \cdot {}_4C_1 \cdot {}_4C_1 = 1,287 \cdot 4 \cdot 4 \cdot 4 \cdot 4 \cdot 4 = 1,317,888$. From this, we subtract the number of straights, flushes, straight flushes, and royal flushes to get: $1,317,888 - 10,200 - 5,108 - 36 - 4 = 1,302,544$. Once again, to find the probability of obtaining this hand, we divide by the total number of possible hands dealt to get $P(\text{only a high card}) = \frac{1,302,540}{2,598,960} = .501177398$, which translates to odds of about 1:1.

Having now considered all of the possible poker hands, the sum of these probabilities should add up to 1, since a probability of 1 or 100 percent is equivalent to certainty. If you are dealt any 5 cards from a deck of 52 cards, they must fall into one of the 10 cases we considered. The sum of their probabilities is shown in table 4.3.

Type of Hand	Frequency	Probability	Odds
Royal flush	4	0.000154%	1:649,739
Straight flush	36	0.00139%	1:72,192
Four of a kind	624	0.0240%	1:4,164
Full house	3,744	0.1441%	1:693
Flush	5,108	0.1965%	1:508
Straight	10,200	0.3925%	1:254
Three of a kind	54,912	2.1128%	1:46.3
Two pair	123,552	4.7539%	1:20.0
One pair	1,098,240	42.2569%	1:1.37
Five different cards	1,302,540	50.1177%	1:0.995
Total:	**2,598,960**	**99.999944 ≈ 100%**	**1:0**

Table 4.3.

Knowing the chances of getting any of these possible poker hands should provide you some guidance about how to play the card game cleverly.

MATHEMATICAL LOGIC OF TIC-TAC-TOE

We often overlook that a part of the study of mathematics is based on logic, and the reverse is also true. This can be seen with the game of tic-tac-toe. Most people are familiar with the basic game of tic-tac-toe (sometimes called "Noughts and Crosses" in Great Britain).[5] As far back as 1300 BCE traces of the game could be found in ancient Egypt. But in the Western world the game seems to have its origins in Rome in the first century BCE, where it was called *Terni Lapilli*. There is also evidence that the game might have originated in ancient Egypt. The first written reference to the name "tick-tack-toe" was a children's game in 1884. The popularity of the game entered the world of computers in 1975 when as a student project at MIT it was demonstrated that a Tinkertoy computer was able to play the game perfectly. This is on display at the Museum of Science in Boston.

Most are familiar with this game, where players alternate turns placing an "X" or an "O" in an empty cell on a 3 × 3 or 9-cell grid as shown in figure 4.1.

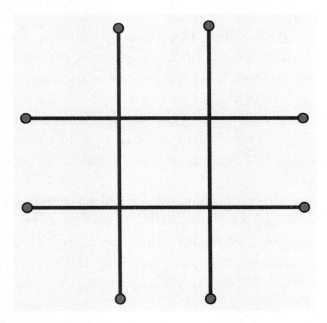

Figure 4.1.

The goal is to get three of your own marks in a straight row, with no intervening spaces or an opponent's mark. Most of us have played this game and developed a strategy that will lead to either a win or a draw. The majority of players usually prefer to go first. Since there are nine cells on the board, going first ensures a chance to place your mark in five cells total, as opposed to the opponent's four cells. Furthermore, most players going first seem to place their mark in the center square. This square is involved in a winning position four times out of eight. Corner squares are involved in a winning position three times out of eight. Sounds easy enough, doesn't it? And yet, the center cell as a first move is not the best approach. The player going first should place his or her X or O in one of the four corner cells. This corner-cell placement is the one that can lead to developing two possibilities for a win. Once these two possibilities have been developed, the opponent cannot block both at once, and a win is assured. If both players are careful and know the strategy, then the very least that can occur is a tie.

Let's consider a strategy used in playing tic-tac-toe. We will have the player using the X go first, and he has three options for his first move. He can place the X in either a corner cell, a middle border cell, or the center cell. This player can force a draw or can win with any of these three starting moves. However, by placing the first move in a corner cell, the second player is limited in his move to avoid losing. The second player, using an O, must respond to the first player's move defensively.

If the first player chooses to place the X in a corner cell, then the second player must place the O in the center cell. If the first player chose to place the X as the initial move in the center cell, then the second player must place his O in a corner cell. If the first player places his initial move by placing the X in a side middle cell, then the second player must place the O in either the center cell, a corner cell next to the O-placed cell, or the side-middle cell opposite the X cell. Any other move by the second player will result in a win by the X player. After these two moves, the second player continuously places Os to block the first player's attempt to get three Xs in a row. Of course, there is always the possibility that the first player, X, might make a careless move that allows the second player, O, to win rather than just to force a draw.

For the second player, placing O, to guarantee a draw, he would need to adhere to the following: If the first player does not place the X in the center cell, then the second player should occupy the center cell with the O, followed by a side middle cell. If the first player places an X in the corner square, and the second player places an O in the opposite corner, that will allow the first player to win if he places the second X in one of the unoccupied corners. If the first player places an X in the corner cell, then the only way the second player can force a tie is to place an O in the center cell and then, as his next move, to place an O in a side-middle cell. By this time, you should be able to notice the various strategies that clever players would use to always end up in a tie. It is when one player makes a careless move the other player will be able to win.

When we analyze the game from a mathematical point of view, there are exactly 138 possible final positions for the game's end. Assuming that the X makes the first move, then 91 positions would have the X as winner; 44 positions would have the O as winner; and 3 positions would have the game end in a draw.

Let's make a simple change in the game, thus creating another game and requiring another strategy. In this new version of the game, three in a row *loses* the game. Does the strategy change? Do you still want to go first? Do you still want to place your X or O in the corner square? A new problem has been presented, and a different strategy must be developed. Play the game with this new goal and see what transpires.

Consider this simulation of the game. Suppose the first player places an X in the center cell. If the second player places an O in a corner cell, then the first player is forced to have two of his marks in a row in more than one position. How can this be avoided? Was this a good first move? Notice that once a player occupies the center cell, his next move *guarantees* two in a row, which turns out to be a bad strategy indeed.

Once you have mastered the original game of tic-tac-toe and can either win or tie almost every time, several thoughts should come to mind about this particular version of the game (in which having three marks in a row causes you to lose). First of all, since there are nine empty cells, you might think it is preferable to go second, since this forces your opponent to place five of his marks on the grid, while you need to place only four of your marks. However, this doesn't always work. Instead, if you go first, place your mark in one of the side-middle cells of the board. These four cells are those least likely to result in three in a row. If you go second, do the same thing. You will quickly see how this strategy can be effective.

Even a strategy game as simple as the original tic-tac-toe affords many opportunities to utilize mathematical skills. Explaining the rationale for making a particular move helps develop your reasoning skills. Communicating these ideas verbally creates an informal dialogue that leads to increasing higher-order thinking and problem-solving skills. Every move requires an informal calculation of the probability of winning or losing the game. Even an opponent's move raises the question of the probability of winning the game.

THE MONTY HALL PROBLEM

It is rare that a mathematics problem would appear on the front page of the *New York Times*, as it did on July 21, 1991. During the year prior to its appearance, it was quite a controversy, generated by Marilyn vos Savant in a question-and-answer column in *Parade* magazine. Mathematicians were arguing about the proper solution to a problem posed as a result of a popular television show, *Let's Make a Deal*. This program was a long-running television game show that featured a problematic situation for a randomly selected audience member to come on stage, be presented with three doors, and be asked to select one. Hopefully the contestant would select the door with the car and not one of the other two doors, each of which had a donkey behind it. There was only one wrinkle in this: After the contestant made her selection, the host, Monty Hall, would expose one of the two donkeys behind one of the unselected doors (leaving two doors still unopened). Then the audience participant would be asked if she wanted to stay with her original selection (not yet revealed) or switch to the other unopened door. At this point, to heighten the suspense, the rest of the audience would shout out "stay" or "switch" with seemingly equal frequency. The question is, What should she do? Does it make a

PROBABILITY, GAMES, AND GAMBLING 155

difference? If so, which is the better strategy to use here (i.e., which provides the greater probability of winning)?

This problem has caused many an argument in academic circles, and it was also a topic of discussion in the *New York Times* and other popular publications as well. John Tierney wrote in the *New York Times* on Sunday, July 21, 1991, that "perhaps it was only an illusion, but for a moment here it seemed that an end might be in sight to the debate raging among mathematicians, readers of *Parade* magazine and fans of the television game show *Let's Make a Deal*. They began arguing last September after Marilyn vos Savant published a puzzle in *Parade*. As readers of her 'Ask Marilyn' column are reminded each week, Ms. vos Savant is listed in the *Guinness Book of World Records Hall of Fame* for 'Highest I.Q.,' but that credential did not impress the public when she answered this question from a reader." She gave the right answer, but still many mathematicians argued.

Let us look at this now step by step. Gradually, the result will become clear.

There are *two donkeys* and *one car* behind these doors.

You must try to get the car. You select Door #3 (Fig. 4.2.)

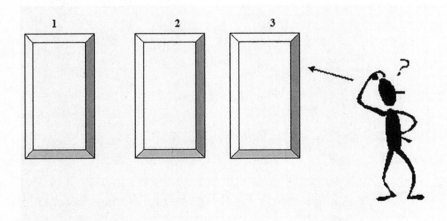

Figure 4.2.

156 THE MATHEMATICS OF EVERYDAY LIFE

Monty Hall opens one of the doors that you *did not* select, and exposes a donkey. (Fig. 4.3.)

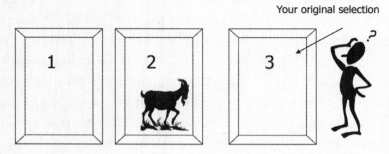

Figure 4.3.

Monty asks: "Do you still want your first-choice door, or do you want to switch to the other closed door"?

To help make a decision, consider an *extreme case*:

Suppose there were one thousand doors instead of just three doors. (See fig. 4.4.)

Figure 4.4.

You choose Door No. 1,000. How likely is it that you chose the right door?

Very unlikely, since the probability of getting the right door is $\frac{1}{1000}$.

How likely is it that the car is behind one of the other doors? (Fig. 4.5.)

Very likely: $\frac{999}{1000}$

PROBABILITY, GAMES, AND GAMBLING   157

Figure 4.5.

These are all "very likely" doors!

Monty Hall now opens all the doors except one (2-999), and shows that each one had a goat.

A "very likely" door is left: Door #1

Which is a better choice?
- Door #1000 ("Very unlikely" door)
- Door #1 ("Very likely" door.)

Figure 4.6.

Monty Hall now opens *all* of the doors (2 – 999) *except one* (say, Door No. 1) and shows that each one had a donkey. (Fig. 4.6.)

A "very likely" door is left: Door No. 1.

We are now ready to answer the question. Which is a better choice:

- Door No. 1,000 ("Very unlikely" door), or
- Door No. 1 ("Very likely" door)?

The answer is now obvious. We ought to select the "very likely" door, which means "switching" is the better strategy for the audience participant to follow.

In the extreme case, it is much easier to see the best strategy than had we tried to analyze the situation with the three doors. The principle is the same in either situation. Here we see a simple solution to a vexing problem that seems to permeate society, even beyond mathematicians. This problem has generated excitement throughout our culture so much so that an entire book has been written on the subject: *The Monty Hall Problem: The Remarkable Story of Math's Most Contentious Brain Teaser*, by Jason Rosenhouse.[6] If you are interested, this book will not only explain the problem, as we did above, but it will also present a variety of variations that lead very nicely to an introduction to probability, a branch of mathematics becoming increasingly more popular in today's technological world. Another form of "gambling" or "gaming" can be seen from the investments viewpoint—our next focus.

BUSINESS APPLICATIONS

The Fibonacci numbers (1, 1, 2, 3, 5, 8, 13, 21, 34, 55, 89, 144, . . .) seem to pop up in the most unexpected places. As we consider the volatile world of the stock market, we find them there as well.

Investor confidence was at an all-time low after the stock market's Great Crash of 1929, whereupon many Americans viewed the stock market as a costly roll of the dice. However, a little-known, albeit successful, accountant and engineer named Ralph Nelson Elliott (1871–1948) decided to pore over decades of stock performance charts and trace trader movement in an attempt to make sense of the crash. As his research progressed, distinct and repetitive zigzagging patterns began to emerge as Elliott studied that ups and downs or, rather, ebb and flow, of market behavior. He dubbed these patterns "waves" and categorized them as either "impulsive" or "corrective" waves that could be measured and used to forecast market behavior.[7]

In a letter to Charles Collins, the publisher of a national market newsletter of the day with an extensive following, the enthusiastic Elliott wrote that there was "a much-needed complement to Dow theory," which he called his "wave theory."[8] This correspondence began in November 1934 when Elliott wrote to Collins about his "discoveries," with the hope that he would find support from Collins. Collins was impressed by the accuracy of Elliott's analysis and invited him to Detroit to explain the process in greater detail. Although Elliott's insistence that all market decisions should be based on wave theory prevented Collins from directly employing Elliott, he did help him to establish an office on Wall Street. Later, in 1938, Collins wrote a booklet, under Elliott's name, titled *The Wave Principle*. Elliott went on to further publicize his theory through various letters and magazines, including the *Financial World* magazine. Then, in 1946, Elliott expanded on "the wave principle" with a book called *Nature's Law: The Secret of the Universe*.[9]

Elliott's famous and influential booklet, *The Wave Principle*, is based on the belief that market behavior should not be seen as random or chaotic but rather as a natural reflection of investor confidence—or lack thereof—(impulsive waves) and self-sustaining market mechanisms (corrective waves). In layman's terms, Elliott believed that the market, like "other things in the universe," moves in predictable cycles once its patterns of behavior have been established and made visible to the trained eye. At this point, it became apparent that Fibonacci numbers determined the pattern.

What may be surprising is just how pervasive Fibonacci numbers are in Elliott's analysis of the stock market. After tracing most bear markets, he found that they move in a series of 2 impulsive waves and 1 corrective wave, for a total of 3 waves. On the other hand, his research showed that a bull market usually jumps upward in 3 impulsive waves and 2 corrective waves for a total of 5 waves. A complete cycle would be the total of these moves, or 8 waves. Recall, the Fibonacci numbers are 1, 1, 2, 3, 5, 8, 13, 21, 34, 55, 89, 144 . . . The Fibonacci sequence's relationship to the wave principle does not end there.

Elliott's wave principle also states that each "major" wave can be further subdivided into "minor" and "intermediate" waves. A regular bear market, where the prices are falling, has 13 intermediate waves. And as now would be expected, a bull market, where the prices are rising, has 21 intermediate waves, for a total of 34. Continuing on with the sequence, there are 55 minor waves in a bear market and 89 in a bull market for a total of 144. Elliott, himself, was not at all surprised to find the Fibonacci sequence while tracing these market waves, as he had always believed that, "the stock market is a creation of man and therefore, reflects human idiosyncrasy." However, it was only when Elliott began to study the relationships between the size of the waves for his second book, *Nature's Law: The Secret of the Universe*, that he realized his new discoveries were based on "a law of Nature known to the designers of the Great Pyramid 'Gizeh,' which may have been constructed 5,000 years ago," which is based on the golden ratio and which, in turn, can be shown to be generated by the ratio of consecutive Fibonacci numbers. Recall that two quantities are in the golden ratio if their ratio is the same as the ratio of their sum to the larger of the two quantities.

Elliott believed that this famous golden ratio along with other Fibonacci-generated ratios could be used to predict stock prices with astounding accuracy. In order to understand how he reached this conclusion, it is important to investigate what these Fibonacci ratios are. Consider the results when dividing Fibonacci numbers by their three immediate successors. Can you see a pattern emerge after the first several calculations? (See table 4.4.)

n	F_n	$\dfrac{F_n}{F_{n+1}}$	$\dfrac{F_n}{F_{n+2}}$	$\dfrac{F_n}{F_{n+3}}$
1	1	$\frac{1}{1} = 1.000000000$	$\frac{1}{2} = 0.500000000$	$\frac{1}{3} \approx 0.333333333$
2	1	$\frac{1}{2} = 0.500000000$	$\frac{1}{3} \approx 0.333333333$	$\frac{1}{5} = 0.200000000$
3	2	$\frac{2}{3} \approx 0.666666667$	$\frac{2}{5} = 0.400000000$	$\frac{2}{8} = 0.250000000$
4	3	$\frac{3}{5} = 0.600000000$	$\frac{3}{8} = 0.375000000$	$\frac{3}{13} \approx 0.230769231$
5	5	$\frac{5}{8} = 0.625000000$	$\frac{5}{13} \approx 0.384615385$	$\frac{5}{21} \approx 0.238095238$
6	8	$\frac{8}{13} \approx 0.615384615$	$\frac{8}{21} \approx 0.380952381$	$\frac{8}{34} \approx 0.235294118$
7	13	$\frac{13}{21} \approx 0.619047619$	$\frac{13}{34} \approx 0.382352941$	$\frac{13}{55} \approx 0.236363636$
8	21	$\frac{21}{34} \approx 0.617647059$	$\frac{21}{55} \approx 0.381818182$	$\frac{21}{89} \approx 0.235955056$
9	34	$\frac{34}{55} \approx 0.618181818$	$\frac{34}{89} \approx 0.382022471$	$\frac{34}{144} \approx 0.236111111$
10	55	$\frac{55}{89} \approx 0.617977528$	$\frac{55}{144} \approx 0.381944444$	$\frac{55}{233} \approx 0.236051502$
11	89	$\frac{89}{144} \approx 0.618055555$	$\frac{89}{233} \approx 0.381974249$	$\frac{89}{377} \approx 0.2360784271$
12	144	$\frac{144}{233} \approx 0.618025751$	$\frac{144}{377} \approx 0.381962865$	$\frac{144}{610} \approx 0.236065574$
13	233	$\frac{233}{377} \approx 0.618037135$	$\frac{233}{610} \approx 0.381967213$	$\frac{233}{987} \approx 0.236068896$

Table 4.4.

It is clear that after the first several rows, the results in each of the preceding rows approximate 0.6180, 0.3820, and the golden ratio, 0.2360. Written as percentages, these numbers translate to: 61.8 percent, 38.2 percent, and 23.6 percent and are called Fibonacci percentages.

(It is interesting to notice a very surprising property: The sum of the second two percentages equals the first (23.60 + 38.20 = 61.80), and the first two percentages (38.20 and 61.80) happen to have a sum of 100 percent.

Elliott's premise, or Fibonacci indicator, is that the ratio, or the proportional relationship, between two waves could be used to indicate stock prices. He found that an initial wave up in price was usually followed by a second wave down, or what market analysts call a "retracement" of the initial surge. Upon closer inspection, these retracements seemed, more often than not, to be Fibonacci percentages of the initial price surge, with 61.8 percent as the highest retracement. Once again, the golden ratio reigns supreme![10]

While many of today's market analysts incorporate wave patterns and retracements to forecast the market, some analysts take Fibonacci numbers a step further. There are market analysts who scan previous dates and results looking for Fibonacci numbers. There might be 34 months or 55 months between a major high or low, or as few as 21 or 13 days between minor fluctuations. A trader might look for repetitive patterns and try to establish a Fibonacci relationship between dates to help forecast the future market.

Before you scoff at the importance of Fibonacci numbers, consider that Elliott, at the age of sixty-seven, and without the aid of any computers, forecast to the *exact day* the end of a bear market decline from 1933 to 1935. As for the golden ratio, consider also the vehicle by which many of these trades are made nowadays: the credit card. It measures 55 mm by 86 mm and is very close to being a golden rectangle (actually, it's 3 mm short in the length from fitting the Fibonacci ratio approximation).

More recently, a book by Robert Fischer called *Fibonacci Applications and Strategies for Traders*[11] expounds on the use of the Fibonacci numbers in investment strategies, and it is a complete, hands-on resource that shows how to measure price and time signals quickly and with accuracy using the logarithmic spiral, and how to act on these analyses. This innovative trading guide first takes a fresh look at the classic principles and applications of the Elliott wave theory, then introduces the Fibonacci sequence, exploring its appearance in the many related fields and then in stock and commodity trading.

Fischer then explains how the series is used to measure equity and commodity price swings and to forecast short- and long-term correction targets. You can learn how to accurately analyze price targets on market extensions, and how, by using the author's logarithmic spirals, to easily develop price and time analyses with some precision. The book claims to enable you to calculate and predict key turning points in the commodity market, analyze business and economic cycles, and discover entry and exit rules that make disciplined trading possible and profitable. As you can see, even today, the Fibonacci numbers continue to fascinate us in the investment field.

MATHEMATICS OF LIFE INSURANCE

Although it might feel macabre, the whole field of life insurance is based on mathematical probability and statistics. Most people who buy life insurance are relatively clueless about how the costs are calculated. We will consider a simple case here just to give you an idea of what lies behind the premiums you pay. Let's begin with the following problem: Out of 200,000 men alive at age forty, 199,100 lived to become age forty-one. What is the probability that a forty-year-old insured man will live at least one year? What is the probability that he will die within one year?

This problem should also begin to make you aware of the applicability of probability theory to life insurance. Life insurance companies must be able to measure the financial risks of insuring particular people. To decide on the premiums, a life insurance company must know how many people in any group are expected to die within a specific time frame. They do this by collecting data about the number of people (from each age group) who died in the past within the time frame being considered. Since the data is collected from a large number of events, the law of large numbers applies. This law states that *with a large number of experiments, the ratio of the number of successes to the number of trials gets very close to the theoretical probability.*

Although it sounds morbid to those of us outside of the industry, life insurance companies construct mortality tables based on past deaths in order to predict the number of people who will die in each age group. Merely as a sample we will use a portion of the Commissioners 1958 Standard Ordinary Mortality Table (table 4.5). To construct this table, a sample of 10 million people was used. Their life span was recorded from birth until age ninety-nine. At each age level, the table records the number of people alive at the start of the year and the number of deaths that occurred during the year. Then the following ratio is computed:

$$\frac{\text{Number of deaths during the year}}{\text{Number of people alive at the start of the year}}$$

This ratio is then converted to deaths per 1,000, which is called the *death rate*. This death rate, as you will see, is crucial in computing the premium that policyholders will pay.

PROBABILITY, GAMES, AND GAMBLING

Age	Number Living	Deaths Each Year	Deaths per 1,000
0	10,000,000	70,800	7.09
1	9,929,200	17,475	1.76
2	9,911,725	15.066	1.52
3	9,896,659	14,449	1.46
4	9,882,210	13,835	1.40
10	9,805,870	11,865	1.21
11	9,794,005	12,047	1.23
12	9,781,958	12,325	1.26
13	9.769,633	12,896	1.32
18	9,698,230	16,390	1.69
25	9,575,636	18,481	1.93
30	9,480,358	20,193	2.13
42	9,173,375	38,253	4.17
43	9,135,122	41,382	4.53
44	9,093,740	44,741	4.92

Table 4.5.

We are now ready to address the following problem.

The problem: **What is the probability that an eighteen-year-old will die, if, out of 6,509 eighteen-year-olds alive at the beginning of the year, 11 died?**

The probability is $\frac{11}{6,509}$. However, life insurance companies prefer to transform this ratio into death rate per 1,000. For general purposes, we will change the fraction $\frac{11}{6,509}$ into $\frac{x}{1,000}$ to set up the following proportion: $\frac{x}{1000} = \frac{\text{Number of deaths during the year}}{\text{Number of people alive at the start of the year}}$, where x = death rate per 1,000. This gives us $\frac{11}{6,509} = \frac{x}{1,000}$, or $x = 1.69$. This means that 1.69 people out of the original 1,000 will have died by the end of the eighteenth year. The insurance companies use this information to calculate the premium they will charge a group of eighteen-year-olds. Suppose there were 1,000 eighteen-year-olds who insured themselves for $1,000 each for one year.

How much would the company have to pay out at the end of the year? If 1.69 people die, the company will pay out $1,690 (1.69 · 1,000 = 1,690). Thus, how much must the company charge each of the 1,000 policyholders? (This does not take into account profit or operating expenses.) The $1,690 divided evenly among 1,000 people equals $1.69 per person.

In the previous discussion, we did not take into consideration the fact that money paid to the company earns interest during the year. So, besides considering the death rate, the interest rate must also be taken into account when calculating the premium. We now need to recall from the previous chapter the concept of compound interest. Let's consider how much money will be on deposit in a bank at the end of the year if you deposit $100 at 5 percent interest. The answer is $100 plus 0.05 (100), or 100 · 1.05, which is $105. If the $105 is kept in the bank another year, what will it amount to? Let's look at the math: $105 + 0.05(105), or $100 · 1.05 · 1.05 or $100 · $(1.05)^2$, which amounts to $110.25. We will then write the general formula using P = original principal, i = rate of interest per period, A = the amount of money at the end of the specified time, and n = the number of years the principal is on deposit. The formula is $A = P(1 + i)^n$.

Consider how much money you would have to deposit now in a bank whose rate of interest is 5 percent, if you wanted $100 to accumulate by one year from now. In the previous example, we saw that $100 grew to $105 in one year's time. This information is used to set up a proportion:

$$\frac{x}{100} = \frac{100}{105} = 0.9524, \text{ and } x = 100(0.9524) = \$95.24.$$

How much would have to be deposited now to accumulate $100 at the end of two years' time?

It would be $\frac{x}{100} = \frac{100}{110.25} = 0.9070$, and $x = \$90.70$. We are now ready to derive a formula for calculating the present value from the formula for compound interest, which we derived above as

$A = P(1 + i)^n$. This present-value formula is $P = \frac{A}{(1+i)^n}$.

We can now return to the original problem of the life insurance company that has to pay out $1,690 at the end of the year to the deceased eighteen-year-old's beneficiaries. What is the present value of $1,690? In other words, how much must the insurance company collect at the beginning of the year so that it can pay out $1,690 at the end of the year? By using the present value formula, we computed that for every $1 the company has to pay, it must collect $0.9524 at the beginning of the year. If the company has to pay $1,690, then it has to collect $1,609.56 in total from its 1,000 eighteen-year-olds (1,690 · 0.9524 = $1,609.56). Thus, each policyholder must contribute a premium of $\frac{\$1,609.56}{1,000} = \$1.60956 \approx \$1.61$.

You may now pose another problem. Suppose another group of 1,000 people aged twenty-five bought policies for one year worth $1,000 apiece (the death benefit is $1,000). According to the mortality table, their death rate is 1.93; this means that 1.93 out of 1,000 twenty-five-year-olds die during their twenty-fifth year. What will the net premium be if the interest rate is 5 percent? Again, the death rate per 1,000 at age twenty-five = 1.93. The amount needed to pay claims = (1.93 · 1,000) = $1,930. The interest factor = $0.9524. The present value of claims due in one year ($1,930 · 0.9524) = $1,838.13, and the number of persons paying premium = 1,000. Therefore, the net premium $\frac{\$1,838.13}{1,000} = 1.83813 \approx \$1.84$. This process may be continued for additional years of insurance.

We have only taken a simple sample to provide some insight into the calculation of life insurance premiums. It is clear that life insurance companies need to earn money and at the same time provide a justifiable premium. Obviously, mathematics plays a key role here. Enough about finance! Let's look at some counterintuitive arithmetic in everyday life.

THE MOST MISUNDERSTOOD AVERAGE

Most unsuspecting readers, when asked to calculate the average speed for a round trip with a "going" average speed of 30 miles per hour and a "returning" average speed of 60 miles per hour would think that their average speed for the entire trip is 45 miles per hour (calculated as $\frac{30+60}{2} = 45$). The first task is to realize that this is the *wrong answer*. Is it fair to consider the two speeds with equal "weight"? After all, the two speeds were achieved for different lengths of time; therefore, we know that they should not get the same weight. We realize that the trip at the slower speed, 30 mph, took twice as long, and therefore ought to get twice the weight in the calculation of the average round-trip speed. This would then bring the calculation to the following: $\frac{30+30+60}{3} = 40$, which happens to be the correct average speed.

Analogously, once again we have a weighted average, which can be seen when we consider the grade a student deserves. Suppose the student scored 100 percent on nine of ten tests in a semester, and on one test scored only 50 percent. Would it be fair to assume that this student's performance for the term should be graded as $\frac{100+50}{2} = 75$ percent? The reaction to this suggestion will tend toward applying appropriate weight to the two scores in consideration. The 100 percent was achieved nine times as often as the 50 percent and therefore ought to get the appropriate weight. Thus, a proper calculation of the student's average ought to be $\frac{9(100)+50}{10} = 95$ percent. This clearly appears to be more just!

We then might ask what happens if the rates to be averaged are not multiples of one another? For the speed problem above, one could find the time "going" and the time "returning" to get the total time, and then, with the total distance, calculate the "total rate," which is, in fact, the average rate.

There is a more efficient way to handle this kind of a problem, and that is the highlight of this section. We are going to introduce a concept called the *harmonic mean*, which is the mean of a harmonic sequence.

PROBABILITY, GAMES, AND GAMBLING

The name *harmonic* may come from the fact that one such harmonic sequence is $\frac{1}{2}, \frac{1}{3}, \frac{1}{4}, \frac{1}{5}, \frac{1}{6}, \frac{1}{7}, \frac{1}{8}$, and if you take guitar strings of these relative lengths and strum them together, a harmonious sound results.

This frequently misunderstood mean (or average) usually causes confusion. To avoid this, once we identify that we need to find the average of rates (i.e., the harmonic mean), we have a simple formula for calculating the correct average, namely, the harmonic mean for rates over the same base. In the above situation, the rates were for the same distance for both parts of the round trip.

The formula for the harmonic mean for two rates a and b is $\frac{2ab}{a+b}$; and for three rates, a, b, and c, the harmonic mean is $\frac{3abc}{ab+bc+ac}$.

You can see the pattern evolving, so that for four rates the harmonic mean is $\frac{4abcd}{abc+abd+acd+bcd}$, and so on.

Applying this to the above speed problem gives us $\frac{2 \cdot 30 \cdot 60}{30+60} = \frac{3,600}{90} = 40$.

Now let's consider the following problem:

The problem: **On Monday, a plane makes a round-trip flight from New York City to Washington and back, with an average speed of 300 miles per hour. The next day, Tuesday, there is a wind of constant speed (50 miles per hour) and direction (blowing from New York City to Washington). With the same speed setting as on Monday, this same plane makes the same round trip on Tuesday. Will the Tuesday trip require more time, less time, or the same time as the Monday trip?**

We should notice that the only thing that has changed is the help and hindrance of the wind (the wind will aid the plane's fight in one direction and hinder it in the opposite direction). All other controllable factors are the same: distances, speed regulation, airplane's conditions, and so on. An expected response is that the two round-trip flights ought to be the same, especially since the same wind is helping and hindering two equal legs of a round-trip flight.

But you should realize that the two legs of the "windy trip" require different amounts of time, which should lead you to the notion that the two speeds of this trip cannot be weighted equally as they were done for different lengths of time. Therefore, the time for each leg should be calculated and then appropriately apportioned to the related speed.

We can use the harmonic mean formula to find the average speed for the windy trip. The harmonic mean is $\frac{2 \cdot 250 \cdot 350}{250+350} = 291.667$, which is slower than the no-wind trip of 300 mph. What a surprise!! Here we have another example of how a little common sense through mathematical thinking lets us navigate our world more accurately.

WHAT WE NEED TO KNOW ABOUT AVERAGES

There are lots of types of averages, which in various ways measure a point of central tendency. There is the usual average, known in mathematics as the *arithmetic mean*. Then there is a *geometric mean*, and a *harmonic mean*, which we just discussed. However, there other measures of central tendency, such as the *mode* of the set of data, which refers to that item or score that comes up most frequently. And then there is the *median*, which is the midpoint of a spectrum of data, regardless of where the strong points or weak points are located. However, before we compare the three types of means mentioned above, we need to dispel the use of the word "average," as it has been used so frequently in sports, particularly in baseball.

Although baseball batting averages seem to permeate sports discussions, few people realize that these "baseball batting averages" aren't really averages in the true sense of the term; they are actually percentages. Most people, especially after trying to explain this concept, will begin to realize that it is not an average in the way they usually define an "average"—the arithmetic mean. It might be good to search the sports section of the local newspaper to find two baseball players who currently have the same batting average but who have achieved their respective averages with a different number of hits. We shall use a hypothetical example here.

Consider two players: David and Lisa, each with a batting average of .667. David achieved his batting average by getting 20 hits for 30 at bats, while Lisa achieved her batting average by getting 2 hits for 3 at bats.

On the next day, both performed equally, getting 1 hit for 2 at bats (for a .500 batting average). You might expect that they then still have the same batting average at the end of the day. But let's examine this further by calculating their respective averages:

David now has 20 + 1 = 21 hits for 30 + 2 = 32 at bats; this gives him a $\frac{21}{32}$ = .656 batting average.

Lisa now has 2 + 1 = 3 hits for 3 + 2 = 5 at bats; this gives her a $\frac{3}{5}$ = .600 batting average.

Surprise! They do not have equal batting averages.

Suppose we consider the next day, during which Lisa performs considerably better than David does. Lisa gets 2 hits for 3 at bats, while David gets 1 hit for 3 at bats. We shall now calculate their respective averages:

David has 21 + 1 = 22 hits for 32 + 3 = 35 at bats, giving him a batting average of $\frac{22}{35}$ = .629.

Lisa has 3 + 2 = 5 hits for 5 + 3 = 8 at bats, giving her a batting average of $\frac{5}{8}$.

Amazingly, despite Lisa's much-superior performance on this day, her batting average, which was the same as David's at the start, is still lower. There is much to be learned from this misuse of the word "average," but, more important, from this discussion you can find an appreciation for the notion of varying weights of items being averaged.

COMPARING MEASURES OF CENTRAL TENDENCY

In mathematics and in statistics we frequently use measures of central tendency—that is, we use various means such as the *arithmetic mean* (in common terms: the average), the *geometric mean*, and the *harmonic mean*, as mentioned above. Our knowledge of them has been traced back to ancient times. As a matter of fact, the historian Iamblichus of Chalcis (ca. 245–325 CE) reported that after a visit to Mesopotamia, Pythagoras (ca. 580/560–ca. 496/480 BCE) brought back to his followers a knowledge of these three measures of central tendency. This may be one reason why today they are often referred to as *Pythagorean means*. We tend to use these means in statistical analyses, but when we inspect them and compare them algebraically and geometrically, some rather-enlightening properties are revealed.

Let's begin by introducing these three means for two values a and b as follows:[12]

- The *arithmetic mean* is: $AM(a, b) = a \:Ⓐ\: b = \frac{a+b}{2}$;
- The *geometric mean* is: $GM(a, b) = a \:Ⓖ\: b = \sqrt{a \cdot b}$; and
- The *harmonic mean* is: $HM(a, b) = a \:Ⓗ\: b = \frac{2}{\frac{1}{a}+\frac{1}{b}} = \frac{2ab}{a+b}$.

For convenience throughout this discussion we will use these representations as we discuss them in greater detail:

The Arithmetic Mean

Before we compare the relative magnitude of these measures of central tendency, or means, we ought to see what they actually represent. The arithmetic mean is simply the commonly used "average" of the data being considered—that is, the sum of the data divided by the number of data items included in the sum. In a simple example, if we want to find the average—or arithmetic mean—of the two values 30 and 60, we take their sum, 90, and divide it by 2 to get 45.

We can also see the arithmetic mean as taking us to the middle of an arithmetic sequence—that is, a sequence with a common difference between terms—such as 2, 4, 6, 8, 10. To get the arithmetic mean within this sequence, we divide the sum by the number of numbers being averaged. Here we have: $\frac{2+4+6+8+10}{5} = \frac{30}{5} = 6$, which, as we expected, happens to be the middle number in the sequence of an odd number of values.

The Harmonic Mean

If we take an arithmetic sequence such as 1, 2, 3, 4, 5, and then take the reciprocals, we have a harmonic sequence: $1, \frac{1}{2}, \frac{1}{3}, \frac{1}{4}, \frac{1}{5}$. We can tie the harmonic mean to the arithmetic mean by simply indicating that the harmonic mean is the reciprocal of the arithmetic mean of the reciprocals of the numbers. We would do this step by step as follows:

To get the harmonic mean $HM(1, 2, 3, 4, 5)$ of a given sequence 1, 2, 3, 4, 5, we first find the arithmetic mean, $AM\left(1, \frac{1}{2}, \frac{1}{3}, \frac{1}{4}, \frac{1}{5}\right)$, of the reciprocals of the sequence, that is,

$$AM\left(1, \frac{1}{2}, \frac{1}{3}, \frac{1}{4}, \frac{1}{5}\right) = \frac{1 + \frac{1}{2} + \frac{1}{3} + \frac{1}{4} + \frac{1}{5}}{5} = \frac{\frac{60+30+20+15+12}{60}}{5} = \frac{\frac{137}{60}}{5} = \frac{137}{300} \ (\approx 0.457).$$

We then take the reciprocal of this value to get the harmonic mean,

$$HM(1, 2, 3, 4, 5) = \frac{300}{137} (\approx 2.19).$$

Another way of looking at the same procedure is the following:

$$HM(1, 2, 3, 4, 5) = \frac{1}{AM\left(1, \frac{1}{2}, \frac{1}{3}, \frac{1}{4}, \frac{1}{5}\right)} = \frac{1}{\frac{1 + \frac{1}{2} + \frac{1}{3} + \frac{1}{4} + \frac{1}{5}}{5}} = \frac{300}{137} \ (\approx 2.19).$$

Where the harmonic mean is particularly useful is to find the average of rates over a common base, as we have seen in the earlier discussion in this chapter.

The Geometric Mean

The geometric mean gets its name from a simple interpretation in geometry. A rather-common application of the geometric mean is obtained when we consider the altitude to the hypotenuse of a right triangle, which, in figure 4.7, is CD, the altitude to the hypotenuse of right triangle ABC.

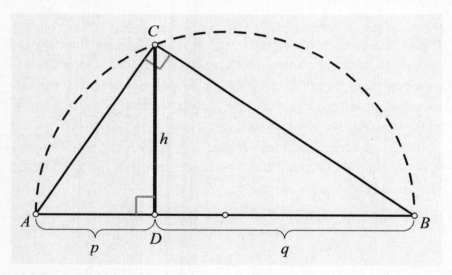

Figure 4.7.

From the triangle similarity $\triangle ADC \sim \triangle CDB$, we get $\frac{AD}{CD} = \frac{CD}{BD}$, or $\frac{p}{h} = \frac{h}{q}$. This then gives us $h = \sqrt{pq}$, which has h as the geometric mean between p and q.

The geometric mean is also the "middle" of a geometric sequence. Take, for example, the geometric sequence, 2, 4, 8, 16, 32. (Note that a geometric sequence is one with a common factor between consecutive terms.) We can extend taking the *square root* when we are averaging *two* items, to taking the *fifth* root when considering *five* items, Therefore, to get the geometric mean of these five numbers, we find the fifth root of their product: $\sqrt[5]{2 \cdot 4 \cdot 8 \cdot 16 \cdot 32} = \sqrt[5]{32,768} = 8$, which is the middle number when the sequence has an odd number of numbers.

Comparing the Three Means Algebraically

Before we present some unusual geometric methods for comparing the magnitude of these three means, we shall show how these three means may be compared in size using simple algebra.

For the two non-negative numbers a and b,

$(a - b)^2 \geq 0$
$a^2 - 2ab + b^2 \geq 0$

Add $4ab$ to both sides:

$a^2 + 2ab + b^2 \geq 4ab$
$(a + b)^2 \geq 4ab$

Take the positive square root of both sides:

$a + b \geq 2\sqrt{ab}$

or $\dfrac{a+b}{2} \geq \sqrt{ab}$

This implies that the *arithmetic mean* of the two numbers a and b is greater than or equal to the *geometric mean*. (Equality is true only if $a = b$.)

Beginning, as we did above, and then continuing along as shown below, we get our next desired result: a comparison of the geometric mean and the harmonic mean.

For the two non-negative numbers a and b,

$$(a - b)^2 \geq 0$$
$$a^2 - 2ab + b^2 \geq 0$$

Add $4ab$ to both sides:

$$a^2 + 2ab + b^2 \geq 4ab$$
$$(a + b)^2 \geq 4ab$$

Multiply both sides by ab:

$$ab(a + b)^2 \geq 4a^2b^2$$

Divide both sides by $(a + b)^2$:

$$ab \geq \frac{4a^2b^2}{(a+b)^2}$$

Take the positive square root of both sides:

$$\sqrt{a \cdot b} \geq \frac{2ab}{a+b}$$

This tells us that the *geometric mean* of the two numbers a and b is greater than or equal to the *harmonic mean*. (Here, equality holds whenever one of these numbers is zero, or if $a = b$).

We can, therefore, conclude that *arithmetic mean* $\geq$ *geometric mean* $\geq$ *harmonic mean*.

Comparing the Three Means Geometrically—
Using a Right-Angled Triangle

The comparison of the three means in terms of their relative size was known to the ancient Greeks, as we find in the writings of Pappus of Alexandria (ca. 290–350 CE). We will now embark on a geometric journey to consider various ways in which the relative sizes of these means can be compared using simple geometric relationships.

In figure 4.8, we have a right triangle with an altitude drawn to the hypotenuse, where the hypotenuse is partitioned by the altitude into two segments of lengths a and b, and $a \leq b$. Here we show the line segments that can represent the three means, and where we can then "see" their relative sizes. (Note, the harmonic mean is denoted with Ⓗ; the geometric mean, with Ⓖ; and the arithmetic mean, with Ⓐ.)

That is, a Ⓗ $b \leq a$ Ⓖ $b \leq a$ Ⓐ b.

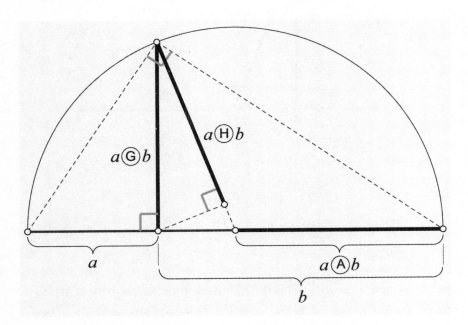

Figure 4.8.

178　THE MATHEMATICS OF EVERYDAY LIFE

To justify our visual observation, we begin by considering figure 4.9, where triangle ABC is inscribed in a circle and AB is a diameter of the circle, implying that ABC is a right triangle. We construct the altitude on AB and join the center of the circle, M, with C. By drawing a line from point D perpendicular to MC, we obtain another right triangle, CED. We notice that CE is a leg of right triangle CED; therefore, $CE \leq CD$. Since the radius, MB or MA, of this circle is longer than the altitude to the hypotenuse of the triangle ABC, we have $CD \leq MB$. Combining these inequalities gives us $CE \leq CD \leq MB$. Our task is to show that these three segments, whose relative lengths we have established, actually represent the three means of a and b as we earlier stated.

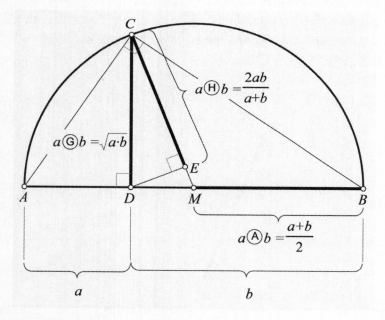

Figure 4.9.

First, we know that AB is the diameter of the circle with center at M and radii $MA = MB = MC = \frac{a+b}{2}$, which is the arithmetic mean between a and b.

To find the geometric mean of a and b, we begin with the altitude, CD, to the hypotenuse of right triangle ABC, which partitions the right triangle into two similar triangles ($\triangle ADC \sim \triangle CDB$); therefore, $\frac{a}{CD} = \frac{CD}{b}$, which leads to $CD^2 = a \cdot b$; thus, $CD = \sqrt{a \cdot b}$, which is the geometric mean of a and b.

From the similar triangles ($\triangle CDM \sim \triangle ECD$) located in right triangle CDM, we can get $CD^2 = MC \cdot CE$, which yields

$$CE = \frac{CD^2}{CM} = \frac{a \cdot b}{\frac{a+b}{2}} = \frac{2ab}{a+b},$$

which is the harmonic mean.

Having now justified how the line segments, which we size-ordered as $CE \leq CD \leq MB$, can represent the various means of a and b, we have therefore shown geometrically that $\frac{2ab}{a+b} \leq \sqrt{a \cdot b} \leq \frac{a+b}{2}$.

Thus, we have considered some of the applications of probability and statistics as they appear in our world and culture. This is merely the beginning of the simple applications that can give you a sense of how mathematics can be used to measure our interests and our environment. Much of this is subsumed by computers, through which we are merely given an answer to a question. But this is what goes on behind the scenes, and we are better off knowing why things are the way they are so that we can make intelligent decisions.

CHAPTER 5

SPORTS AND GAMES— EXPLAINED MATHEMATICALLY

When you engage in sports, typically, the farthest thing from your mind is using mathematics. You need to concentrate on the skills and strategies for winning the game. However, if you become aware of some of the underlying aspects of how mathematics can optimize sports-related moves, you have a distinct advantage over your opponent. For example, you can fine-tune the arc in which to throw a basketball so that it will best reach the hoop. Or you can better determine where a billiard ball should hit the cushion to reach its desired goal, which can prove quite helpful to winning the game. Even physical attributes of some toys can be explained mathematically. In this chapter, we will reveal to you some of these aspects of mathematical applications in sports and games.

THE BEST ANGLE TO THROW A BALL

Throwing or hitting a ball as far as possible is an important skill in many sports, including baseball, football, and golf. In some sports, this is actually the only thing that matters. This is the case most notably for shot put and hammer throwing, which are two sports that belong to the oldest competitions of the modern Olympic Games. Their origins date back at least to the Middle Ages, when soldiers hurled cannonballs as a competitive challenge. The other two throwing events in track-and-field competitions, the discus throw and the javelin throw, were even among the events of the original Olympic Games, which were held in Olympia in ancient Greece throughout classical antiquity—from 776 BCE to the fourth century CE.[1]

Regardless of whether you want to hit a home run, play a drive shot toward the green, or kick a field goal from a great distance, being aware of some basic mathematical properties of ballistic trajectories will be helpful in improving your technique. A ballistic trajectory is the path that a thrown ball or a launched missile without propulsion will take under the action of gravity, if aerodynamic drag is neglected. Although a thrown football, for instance, will not exactly follow a ballistic trajectory because of aerodynamic drag, it is still a fairly good approximation for many purposes. When you throw a ball, there are basically two parameters of the trajectory that you can control: the initial speed and the initial angle, measured from the horizontal. We will not consider spin, since it would have no effect on the trajectory, given that air resistance is not being taken into account. Moreover, even with aerodynamic drag, an additional spinning motion of the ball would not increase the range anyway; only the shape of the trajectory would change. A ballistic (drag-free) trajectory, however, is uniquely determined by the initial speed and the angle at which the so-called projectile is released. Of course, a higher initial speed will also increase the range, but what about the angle? Experience tells us that neither a nearly horizontal path nor a nearly vertical path will be a good choice for maximizing the flight distance. Figure 5.1 shows ballistic trajectories for fixed initial speed but different launch angles. Each curve represents the path a projectile would take for a certain launch angle. These curves are parabolas, which are the common curves presented in high-school algebra.

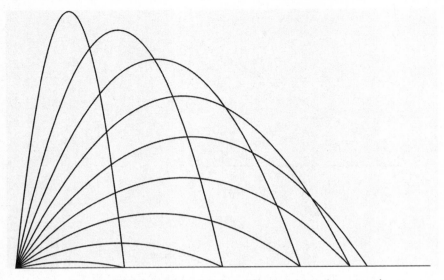

Figure 5.1. Ballistic trajectories for different launching angles.

Clearly, the optimal angle is somewhere "in the middle" between 0° and 90°. You may have been told in physics class that the optimum angle is, in fact, exactly 45°. We will help justify this with the aid of one of the famous fields of mathematics: trigonometry. The initial velocity (a quantity having both magnitude and direction) can be represented by a vector, **V**, that we decompose into a horizontal component V_x, and a vertical component, V_y (see fig. 5.2a). Denoting the launch angle by α, we then have $V_x = V \cdot \cos\alpha$ and $V_y = V \cdot \sin\alpha$, where V is the length of the vector **V**.

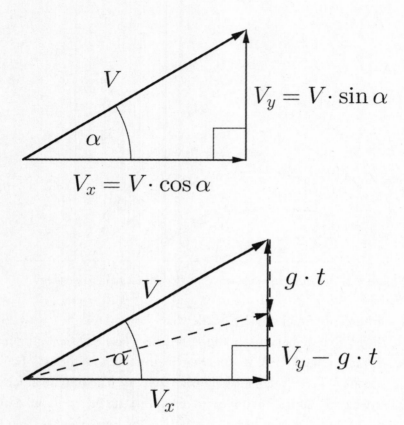

Figure 5.2.

Since we ignore aerodynamic drag, gravitation is the only force acting on the ball during the flight. It pulls the projectile toward the center of the earth with a constant acceleration, $g \approx 9.81 \frac{m}{s^2}$, opposite to the y-axis in our coordinate system. Thus, if you launch the ball at time $t = 0$, the vertical component V_y will be diminished by the amount $g \cdot t$ at time t (see fig. 5.2b). The projectile ascends as long as $g \cdot t$ is smaller than its initial vertical speed, V_y. It reaches the highest point of the trajectory when $V_y = g \cdot t$, implying $t = \frac{V \cdot \sin \alpha}{g}$. From here on, that is, for $t > \frac{V \cdot \sin \alpha}{g}$, gravity dominates and the ball approaches the ground. Since parabolas are always symmetric, we can easily compute the total flight time, T. It must be twice the time at which the maximal height occurs; that is, $T = 2\frac{V \cdot \sin\alpha}{g}$. Hence the longest flight time would be obtained for $\alpha = 90°$, but as this would correspond to launching the projectile vertically upward, no horizontal distance would be gained. So, how can we calculate the distance the projectile travels before it hits the ground? In our drag-free model, the horizontal speed will not change during the flight, since there is no force acting in x-direction. The total horizontal distance traveled, D, is therefore, $D = V_x \cdot T = V\cos\alpha \cdot T = \frac{2 \cdot V^2}{g} \cdot \cos\alpha \cdot \sin\alpha$, where we have substituted $2\frac{V \cdot \sin\alpha}{g}$ for T. To find the optimum angle, we only have to determine for which value of α the horizontal distance, D, attains a maximum, that is, which value of α maximizes the product $\cos\alpha \cdot \sin\alpha$. This can be seen quite easily by invoking a geometrical picture for this problem. Consider a right triangle inscribed in a quarter circle of radius 1, as we show in figure 5.3.

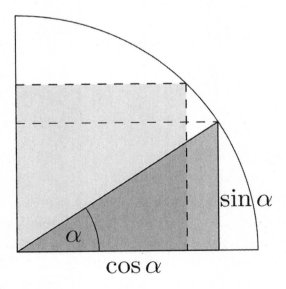

Figure 5.3.

Since the length of the hypotenuse is 1, the sides of the inscribed triangle are of length cosα and sinα, respectively. The product cosα · sinα thus represents the area of a rectangle inscribed in a quarter circle. Maximizing this product means maximizing the area of the rectangle. Clearly, the inscribed rectangle with the largest area is a square, so the optimum angle must be $\alpha = 45°$.

Until now, we have assumed that the projectile is launched directly from the ground, that is, its trajectory starts (and ends) at zero height. Of course, this is not a very realistic scenario for sports moves that involve throwing (rather than, say, kicking). When we throw a ball, we usually stand upright, so the trajectory of the ball will not start on ground level. Taking this into account, the optimal launch angle turns out to be slightly less than 45°. To see this, consider a ball that is released at the point P with an initial speed V, as we show it in figure 5.4.

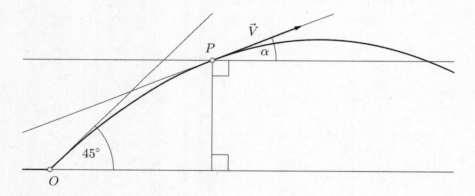

Figure 5.4.

How can we decide whether a trajectory starting at this point with speed V is optimal in the sense that the horizontal distance traveled will be larger than for any other ballistic curve with the same initial speed? We know that if the ball would be launched from ground level, the optimum angle would be exactly 45°. If we can find a parabolic path starting from zero height at an angle of 45° and passing through the point P with speed V, then the angle α of this curve at point P will represent the optimum release angle for a trajectory starting at position P. As can be seen in figure 5.4, this angle obviously measures less than 45°. However, calculating its value for a given initial height above ground level and a given initial speed would require us to delve substantially deeper into trigonometry, so we will only present the result for a typical situation, namely, shot-putting. Here, the idealized zero-drag model is well justified, since the considerable mass of the shot renders air resistance negligible. In shot-putting, the optimum release angle for a typical release height, and initial speed for a world-class throw, would be 42°.

Of course, even for world-class athletes it is impossible to control the release angle this accurately. Moreover, the structure of the human body favors certain characteristics, so meeting the theoretically best release angle might come at the cost of lower initial speed. However, most non-professionals would tend to release a ball or a javelin at a significantly lower angle than 42°. Therefore, if you are active in any of the sports mentioned earlier, then the mathematical considerations presented in this section might compel you to experiment with your release angle, potentially improving your range.

OPTIMIZING YOUR SHOT AT SOCCER

Naturally, there is a good amount of natural talent required to play the game of soccer. When attempting to take a shot on net, there are times when a player will take the ball along the sideline as he runs toward the goal. The question is, At which point along the sideline would the player have the best advantage to kick for a goal? Once again, with mathematics we can provide information as to which point along the sidelines a kicker would have the largest angle to shoot a goal. Of course, during the game such calculations are not possible. Yet it is interesting to see

188   THE MATHEMATICS OF EVERYDAY LIFE

that there is a point along the sidelines from which a kicker has the largest angle to shoot a goal, thereby providing him the optimal position for that attempt. Let's consider a portion of a soccer field as shown in figure 5.5. In this diagram, the net is indicated by points P_1 and P_2; the sidelines are the parallel lines AB and CD; and there are four sample points along AB indicating different locations at which the player can try to aim for a goal. Assume that the player is running along the sideline marked AB and while doing so must determine the point at which his chance of shooting a goal is greatest, in other words, at what point the angle to the goal is largest.

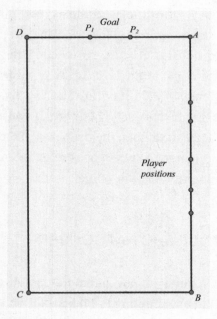

Figure 5.5.

In figure 5.6 we see the various angles that he would have to select from. Which of these would be the largest? And how can we determine the point on the sideline that determines this optimal angle?

SPORTS AND GAMES **189**

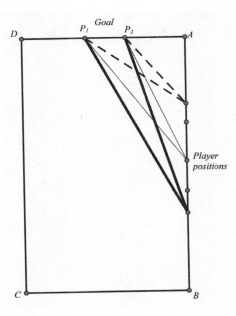

Figure 5.6.

In order to determine the point at which the angle is the largest, we would rely on theorems that were an integral part of the high-school geometry course. That is, we recall that an inscribed angle of a circle is measured by one-half its intercepted arc, and that an exterior angle formed by two secants (that is, lines intersecting the circle in two points and extending beyond the circle) has a measure one-half the difference of the intercepted arcs.

Let's see how this helps us determine the optimum point along the sideline. We notice in figure 5.7 that angle P_1XP_2 is an inscribed angle and is one-half of the measure of arc P_1P_2. Whereas the angles P_1YP_2 and P_1ZP_2 each have a measure that is less than the measure of arc P_1P_2, since they are angles formed by two secants, which have measures equal to one-half the *difference* of their intercepted arcs, that is, one-half arc P_1P_2 *minus* the other intercepted arcs, respectively.

Therefore, the optimal point at which the angle is the greatest is the point at which the circle containing points P_1 and P_2 is tangent to the sideline, which is shown as point X in figure 5.7.

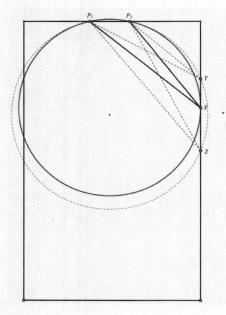

Figure 5.7.

What remains for us now is to determine how we construct a circle that contains two given points (in this case, P_1 and P_2) and is tangent to a given line (in this case, YXZ). To do this, we will work backward from the end product to see what relationships exist, and then construct it going forward.

We will consider the construction using figure 5.8, in which the circle containing the two points P_1 and P_2 is tangent to the line L. If the solution were available, we would see that the line containing chord P_1P_2 meets the given line at a point, A, which is an external point from which a tangent and secant, AT_1 and AP_2P_1, respectively, are drawn to circle O_1.

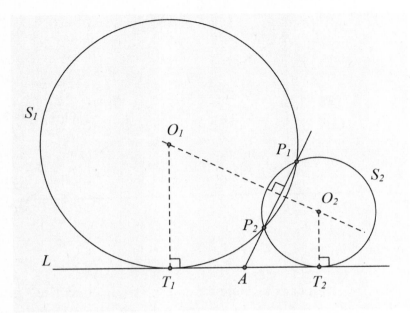

Figure 5.8.

We recall that in such a situation the tangent length is the mean proportional between the length of the whole secant and the length of its external segment. To construct the mean proportional between two given line segments, we merely construct the altitude to the hypotenuse of a right triangle, where the foot of the altitude separates the hypotenuse into the two segments for which the altitude is then the mean proportional. This can be seen in figure 5.9, where AC is the mean proportional between AP_1 and AP_2. This is written as $\frac{AP_1}{AC} = \frac{AC}{AP_2}$.

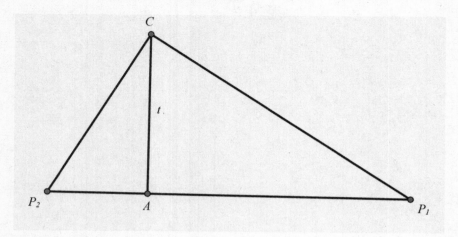

Figure 5.9.

Now we revert back to figure 5.8 and let P_1P_2 intersect line L at A, and construct the mean proportion, AC, between AP_1 and AP_2 (using the right triangle in figure 5.9), and mark along L, on both sides of A, the points T_1 and T_2 so that $AT_1 = AT_2 = AC$. Thus, T_1 and T_2 are tangent points of the required circles on line L. The centers of these circles are on *both* the perpendicular bisector of P_1P_2 and they are also on the perpendicular to L at T_1 and T_2. Thus, we have found a method of constructing the circle that would be tangent to the sideline and also contain the endpoints of the goal.

Although this discussion of locating the optimal point from which to shoot a goal along the sideline is a bit complicated, it demonstrates how simple high-school geometry can answer a question that a soccer player might ask himself when running down the sideline.

A GAME OF ANGLES

"Tennis is a game of angles. You never have time to figure the angles. It's practiced to the extent that it becomes an instinct. You just know where to put the ball. You just feel it. It has been computed into your brain so many times—it is there."[2] This is a quote from Billie Jean King (1943–), one of the most successful tennis players of all time. She won thirty-nine Grand Slam titles, including twelve singles, and is a former World No. 1 player, and in 2006, the USTA National Tennis Center in New York City was renamed the USTA Billie Jean King National Tennis Center in her honor. Undoubtedly, she knows what she is talking about when she refers to tennis as a "game of angles." For professional tennis players, mastering the geometric rules of tennis without having to think about them is no less important than technique, hitting power, and physical strength. Many amateur players underestimate the significance of basic geometric considerations for their success on the court. Tennis tactics and patterns of play are to a very large extent dictated by a geometrical analysis of the position of the players in relation to the court and the possible paths of the ball. "Knowing where to put the ball," that is, understanding and following the geometric rules of a winning strategy, is essential for making the most of your abilities in this sport. We will use geometry from the school curriculum to reveal some of these rules.

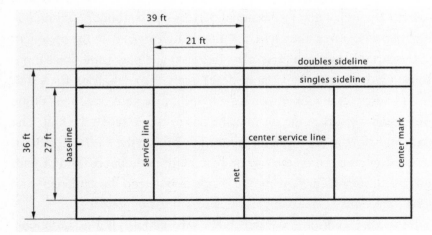

Figure 5.10.

A tennis court is a rectangle that is 78 feet long and 36 feet wide, placed on flat ground with additional clear space outside its boundaries. (See fig. 5.10.) The length is divided in half by a net. The full width of the rectangle is only used in doubles matches. The singles sidelines together with the baselines define the playable area for singles matches (a rectangle of length 78 feet and width 27 feet). The four rectangles around the center of the court are called the *service boxes*. Every point in tennis starts with a *serve*, for which the server must stand behind the baseline, between the center mark and a sideline. For a legal serve, the ball must travel over the net (without touching it) and land in the diagonally opposite service box. The receiver must hit the ball before it has bounced twice and return it onto the other player's half of the court (if the receiver fails, the point goes to the server). A successful return starts a *rally*, in which the players alternately hit the ball across the net. The task is simply to hit one more legal stroke than the opponent. To achieve this goal, there are basically two different strategies you may pursue. A defensive style focuses on returning every ball and waiting for the opponent to make a mistake, while an offensive style means to avoid long rallies by trying to hit aggressive shots that the opponent is unable to return. A legal shot that is not returned by the opponent is called a *winner*. Although many professional players can be roughly fit into one of the two categories, they will adapt their style of play to the opponent and the surface (hard material, clay, or grass) and play sometimes more offensively and sometimes more defensively, depending on the situation. If you go for a winner, you have to minimize your opponent's time to react and maximize the distance he has to run to reach the ball. The opponent's time to react can be reduced by hitting the ball early after it bounces (by taking it on the rise or even before it bounces) and/or with high speed. Key to increasing the distance between the opponent and the ball are shots at sharp angles, away from the opponent.

Before we continue, we will make two simplifying assumptions. First, we assume that the projection of the ball's trajectory onto the court is always a straight line; that is, we neglect possible sidespin of the ball. Second, we will only consider shots that land between the service line and the baseline. Hitting shorter is not a good idea anyway, since it would give your opponent a good opportunity to attack, unless you play a perfect *drop shot*. For a drop shot, you must hit the ball softy and with backspin (rotating backward), so that it barely clears the net and then "drops" on the opponent's half of the court, bouncing low. A good drop shot requires great touch. It is most effective if the opponent is far away from the net and is taken by surprise. However, most of the time it is best to hit the ball "deep," that is, close to the opponent's baseline, keeping the opponent in a defensive position.

The first basic rule for an offensive play is to move closer into the court, whenever it is possible. As soon as your opponent hits short, you should move in and take the ball on the rise. This will reduce your opponent's time to react. But, most important, the closer you move in, the steeper the angles you can create. This is illustrated in figure 5.11. If you are in a neutral position, slightly behind the center of your own baseline (position A), there is a much smaller range of angles you can hit without too much risk than when you take a few steps forward into the court before hitting the ball (position B). Trying to hit a winner from behind the baseline is most of the time not a good idea. From there, you cannot create steep angles. Moreover, the ball will travel longer and give the opponent more time to react and reach the ball.

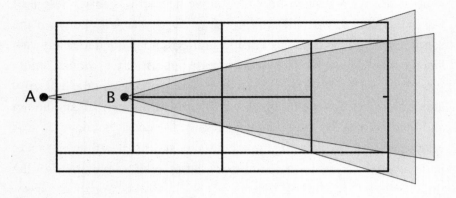

Figure 5.11.

If you are in a position closer to one of the sidelines than to the center line, it is often better to hit the ball *crosscourt*, that is, diagonally, than along the sideline, or "down the line" in tennis jargon. The tennis net is about 15 percent lower at the center than it is at the ends. In addition, by the Pythagorean theorem, the length of the diagonal of a singles court is $\sqrt{78^2 + 27^2} \approx 82.5$ feet, which is more than 5 percent longer than down the line. (See fig. 5.10.) Cross-court shots make the ball pass over the lower portion of the net and along the diagonal of the court (see fig. 5.12). It is safer to hit the ball crosscourt than down the line, since the chances that the ball will hit the net or land outside the playing area are considerably smaller. Thus, crosscourt shots permit a wider range of tolerance, in both height and length of the shot. Shots down the line are generally riskier, since they are more likely to go wide or into the net. However, there is also another reason why it is dangerous to hit the ball down the line in a groundstroke rally (with both players behind the baseline). If you hit the ball down the sideline and your opponent is quick enough to set up a controlled return, you open up the court for your opponent and put him in a geometrically advantageous position. (See fig. 5.12, diagram on the left.) Then, to cover the court against your opponent's potential returns, you have to run a long distance to the middle of his range of angles. If you hit crosscourt instead, you are already at or close to a position from which you can cover most of the possible shots from your opponent. (See fig. 5.12, diagram on the right.) Since you don't have to move very far along the baseline to get in a good position to await your opponent's return, you save energy and you will have more time to set up your next shot, a fact that is compounded further because your shot has a longer distance to travel along the diagonal path.

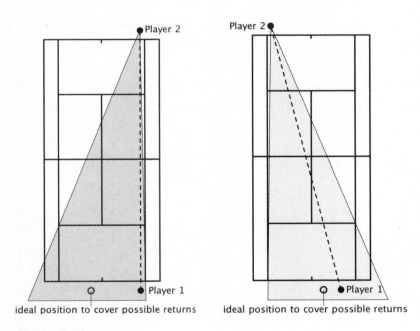

Figure 5.12.

Many amateur players run back to the center mark of the baseline after each stroke, because they think that this is the favorable position to cover as much of the court as possible against their opponent's potential return shots. Yet this is not always true; it depends on the opponent's position on the court. Logically, the best position during a rally is slightly behind the baseline and on an imaginary line that bisects the angle representing the opponent's range of possible shots, indicated by the shaded triangles in figure 5.13. The lines OA and OB represent the broadest returns the opponent may hit. To find your optimal position, you have to bisect angle AOB and place yourself exactly on this bisector, slightly behind the baseline. In figure 5.13, we indicate the optimal spots for different positions of the opponent. Of course, it is impossible to mentally bisect the angle formed by all possible returns in real time, during the rally on the court. But if you take a closer look at the diagrams in figure 5.13, you will find that in all three situations the bisector runs very close to the T on your side of the court, that is, the point of intersection of the service line and the center service line. From this, you can derive a "rule of thumb" to quickly place yourself at least very close to the ideal spot. You just have to visualize an imaginary line from the opponent's point of contact through the T on your side of the court. If you stand on this imaginary line, you optimize your chances to get the next shot.

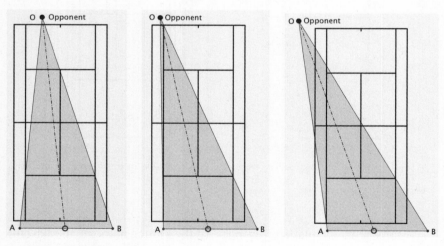

Figure 5.13.

Finally, we want to apply geometry to the situation in which one of the players stands at the net and hits the ball before it bounces on the ground. A shot hit before the ball bounces on your side of the court is called a *volley*. When your opponent is in a defensive position (far behind the baseline), moving close to the net and preparing for a volley makes sense for two reasons: (1), you dramatically reduce the opponent's time to react, and (2), a position at the net increases your range of possible shot angles. From there, you can even attempt a winner by placing the ball only a few feet behind the net and close to the sideline (a shot called a *drop volley*). The risk in approaching the net is to receive a *passing shot* (or *pass*) from your opponent, that is, a shot that passes by the player at the net. Therefore, you should initiate the net attack with an aggressive shot close to the baseline to make it hard for the opponent to set up a passing shot. You then immediately move to the net, following the direction of your last shot. This ensures that you get good coverage for the possible returns your opponent might attempt. The *T*-rule is useless at the net, and it wouldn't apply anyway, since the opponent is forced to risk a passing shot. Contrary to groundstrokes, it is most of the time better to hit volleys down the sideline and as short as possible (and not deep crosscourt). While this may not be very obvious from the perspective of a player on the court, viewing the court from above and using geometry reveals the power of this strategy. In figure 5.14, we show three different options to hit the volley from a position close to the net and in the middle of a service box. The opponent has to hit the ball with sufficient speed to pass the player at the net. Very steep angles are almost impossible to hit with high speed, since the playable court "becomes too short" in these directions. This reduces the possible directions for reasonable shots.

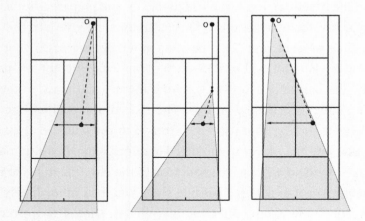

Figure 5.14.

In the diagram on the left in figure 5.14, the volley is placed deep down the line. Although the angle of possible returns is not very large, the player at the net has to be very quick to cover the distance the range of possible returns demands. We have indicated this distance by arrows. The diagram on the right shows the situation after a volley that has been hit crosscourt and close to the baseline. Here the position of the net player is even worse, since it is almost impossible to run to the left sideline fast enough to cover the required distance. The opponent will happily embrace this invitation for a passing shot down the line. The best choice is to place the volley short down the line, as shown in the diagram in the middle. The opponent has to run a long distance to get to the ball and will have to hit it close the ground and near the net. It will be hard to lift the ball over the net with sufficiently high speed, and even then the player on the net has a very good coverage of all possible angles.

It must be said that you need a huge amount of practice to make the right decisions during a rally on the tennis court and finally have the best shot selection "computed in your brain," but all guiding principles for successful patterns of play are actually nothing more than applying geometric concepts. Understanding tennis geometry will give you a significant advantage over your opponents and enable you to beat them even if they hit harder or run faster than you.

PLAYING BILLIARDS CLEVERLY

Those who play the game of billiards, and play it well, have a natural talent for knowing how to hit a ball against the cushion so that it bounces off and lands at a predetermined spot. However, using mathematics, we can find that desired point rather simply. But before we employ a mathematical strategy that will justify the technique, let's discuss the physical aspects involved.

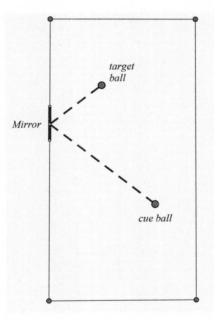

Figure 5.15.

Imagine that you have the cue ball located somewhere on the table as shown in figure 5.15. Your objective is to hit the cue ball against the side cushion and have it ricochet so that it would hit the target ball, also shown in figure 5.15. The technique is very simple. All we need to do is to place a mirror against the side cushion and aim the cue ball at the mirror, shooting it to the point at which the target ball is seen in the mirror. Of course, once you have marked that point, remove the mirror (so that you don't break it) and aim the ball to that point.

The explanation that justifies this technique uses a concept from geometric transformations. In this case, the transformation is a *reflection*. To reflect a given point in the given line, we simply locate the new point along the perpendicular to the given line from the given point and at an equal distance from the line (but on the other side of the line), as shown in figure 5.16.

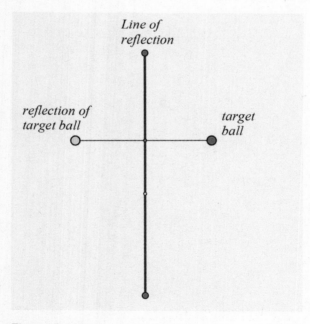

Figure 5.16.

We will now reflect the target ball in the side cushion, which is geometrically equivalent to seeing the reflection of the target ball in the mirror that was placed against the side cushion. We then aim the ball at the mirror-reflection of the target ball with the straight-line segment shown in figure 5.17. The ball will then ricochet off the cushion at the point P and hit the target ball as desired.

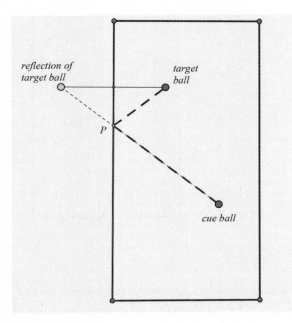

Figure 5.17.

We can easily show that this is the shortest distance from the cue ball to the cushion to the target ball. This can be done by choosing any other point Q on the cushion, and comparing the distance traveled from the cue ball's starting point to the point Q and then to the target ball with the distance of the ball's previous path through point P. (See fig. 5.18.) The path from point C to P to T, or, as stated geometrically, $CP + PT$ is equal to $CP + PR$, since triangle TPR is isosceles. Similarly, $CQ + QT = CQ + QR$. However, we know that the shortest distance between two points is a straight line, therefore, $CP + PR < CQ + QR$. Or, stated verbally, $CP + PR$ is shorter than any other path from the cue ball to the cushion to the target ball.

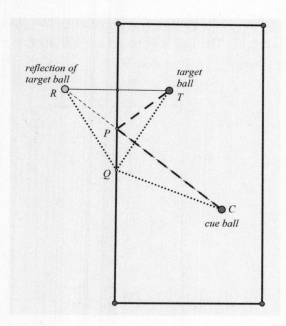

Figure 5.18.

For those who wish to take the billiard example one step further, namely, by hitting two cushions before hitting the target ball, we would need to reflect the target ball in the two cushions and then follow that path, as shown with dashed lines in figure 5.19. That is, first reflect the target ball in one cushion, then reflect this reflection of the target ball in the extension of the other cushion, and then determine the points A and B noted in figure 5.19.

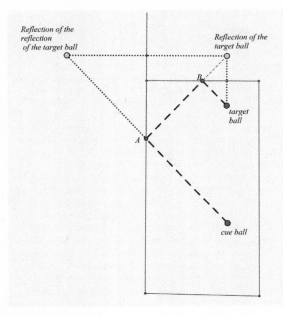

Figure 5.19.

The same effect can be obtained by reflecting each of the two balls in their respective cushions and then joining the two reflections to determine the points *A* and *B*, which are the points of contact on the cushions—as shown in figure 5.20.

206 THE MATHEMATICS OF EVERYDAY LIFE

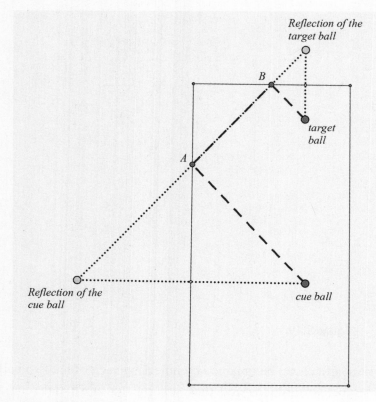

Figure 5.20.

Although the geometric procedure we are using may not be practical to use during the billiard game, it does justify how using one or two mirrors can determine the points of reflection, which are the points that the cue ball would need to hit in order to reach the desired target ball.

MATHEMATICS ON A BICYCLE

With the increased proliferation of bicycles in our society today—especially in big cities—it would be good to see how mathematics can make us better understand the gear selection. Many bicycles have multiple gears and, therefore, also many gear ratios. A shifting mechanism allows us to select an appropriate gear ratio for efficiency or comfort under the prevailing circumstances. The bicycles that we shall consider have two wheels of equal diameters and derailleur gears with one, two, or three sprocket wheels (wheels with teeth or cogs that mesh with the bike's chain) in the front and a sprocket wheel cluster on the rear wheel. The latter typically consists of five or more sprocket wheels, depending on the type of the bicycle. On the rear wheel, the largest sprocket wheel is closest to the spokes, with the rest gradually getting smaller. (See fig. 5.21.)

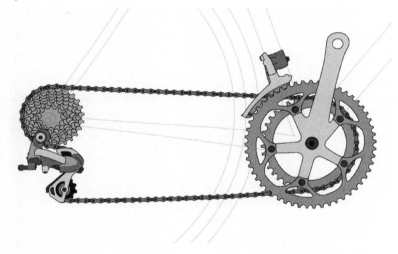

Figure 5.21. Derailleur Bicycle Drivetrain. (Image from Wikimedia Creative Commons, author: Keithonearth; licensed under CC BY-SA 3.0.)

The sprocket wheels in the front are also called *chainrings*. They are attached to the cranks to which the pedals attach. The gearing (that is, the connection of sprocket wheels by a chain) is obtained by moving the chain from one sprocket to the other by means of a derailleur, a device that lifts the driving chain from one sprocket wheel to another of a different size.

Modern racing bikes often have 10-speed or 11-speed cassettes (sprocket clusters) and 2 chainrings in the front. Thus, there are up to 22 different gears possible. Mountain bikes, on the other hand, usually have even three chainrings in the front and, accordingly, up to 33 different gears. This is just a theoretical maximum, since some gears would result in a very diagonal chain alignment and should not be used, as they would cause excessive chain wear. Bicycles for off-road cycling need very low gears for steep grades with poor traction. That's why mountain bikes have a third, particularly small, chainring in the front.

Let's now examine closely the basic setup. There exists a front and rear sprocket wheel with teeth set in gear by a connecting chain. The numbers of teeth on the front and rear sprocket wheels are important. Suppose the front sprocket wheel has 40 teeth and the rear sprocket wheel has 20 teeth; the ratio would then be $\frac{40}{20}$, or 2. This means that the rear sprocket wheel turns twice every time the front sprocket wheel turns once. But the rear sprocket wheel is attached to the bicycle wheel, and so the bicycle wheels will turn twice as well, when the pedal makes one revolution. The distance traveled during one complete revolution of the pedals also depends on the diameter of the drive wheel. Thus, including the bicycle wheel in our consideration, the relevant quantity is the gear ratio, measured in gear inches. It is defined as the product of the diameter of the drive wheel in inches and the ratio between the number of teeth in the front chainring and the number of teeth in rear sprocket. The result is usually rounded to the nearest whole number:

$$\text{gear inches} = \text{bicycle wheel diameter} \cdot \frac{\text{number of teeth in the front chainring}}{\text{number of teeth in the rear sprocket}}$$

Assuming that the diameter of the wheel (including the tire) is 27 inches, we would have $2 \cdot 27'' = 54$ gear inches for a 40-tooth chainring and a 20-tooth sprocket. The number generated here can range from 20 (very low gear) to 125 (very high gear), which provides a comparison between gears, which bicycle experts and racers use to fine-tune the gearing to better suit their level of fitness, their muscle strength, and the expected use. If you multiply the gear inches obtained in the formula above by $\pi \approx 3.14$, you get the distance traveled forward with each turn of the pedals, called "meters of development." (Recall: circumference = $\pi \cdot$ diameter.) A higher gear is harder to pedal than a lower gear, since the distance traveled during one revolution of the pedals is longer, and, therefore, the work performed by the rider with each turn of the pedals must be greater.

For example, a rider using a front-sprocket wheel with 46 teeth and a 16-tooth wheel in the rear along with a 27″ wheel gets a gear ratio of $77.625 \approx 78$. Another rider using a 50-tooth front-sprocket wheel and 16-tooth sprocket wheel in the rear gets a gear ratio of $84.375 \approx 84$, which would be harder to pedal than a 78-gear ratio. The rider with the 78-gear ratio goes approximately 245 inches (or 6.2 meters) forward for each turn of the pedals; the rider with the 84-gear ratio goes approximately 264 inches (or 6.7 meters) forward for each complete turn of the pedals. Hence, the increased difficulty in pedaling is returned in greater distance per pedal revolution.

Now let us examine applications of various gearing ratios to the average rider. Suppose Charlie was riding comfortably on level ground in a 78-gear ratio, and then he came to a rather-steep hill. Should he switch to a higher or lower gear ratio? Your reasoning should be as follows: If Charlie switches to an 84-gear ratio, he will go 264 inches forward for each turn of the pedals. This requires a certain amount of work. To overcome the effects of gravity to get up the hill requires additional energy. So, Charlie would probably end up walking his bicycle. If Charlie had switched to a lower gear, he would use less energy to turn the pedals, and the additional energy required to climb the hill would make his gearing feel about the same as the 78-gear ratio. So, the answer is to switch to a lower-number gear ratio. Charlie will have to turn the pedal more revolutions to scale the hill—more than if he had chosen the 84-gear ratio, and more than if he stayed in the 78-gear ratio. Remember, his gearing only feels like 78 because of the hill.

The work performed while riding a bicycle from A to B is independent of the chosen gear, but a rider can adapt the gear to the slope of the road and to her or his physical condition. The trade-off for reducing the pedaling force by switching to a lower gear is a higher number of pedal revolutions that are necessary to get from point A to point B. However, the force exerted on the pedals and the total number of revolutions are not the only relevant criteria. A cyclist's legs produce power optimally within a narrow pedaling speed range, or cadence. The cadence is the number of pedal revolutions per minute (rpm). For a bicycle to travel at the same speed (or in the same time from A to B), using a lower gear requires the rider to pedal at a faster cadence, but with less force. Conversely, a higher gear provides a higher speed for a given cadence, but it requires the rider to exert greater force. For professional cyclists, choosing the right gear is essentially a matter of finding the optimal balance between the cadence and the pedaling force. In a bicycle race, for instance, a stage of the Tour de France, the cyclists will mostly choose the gear such that the cadence always remains in their preferred range. The preferred cadence varies individually and is somewhere between 70 and 110 rpm for most cyclists. Table 5.1 shows the bicycle's speed for different gears and cadences.

Table 5.1.

Of course, there are circumstances in which the preferred cadence corridor must be temporarily left. This is typically the case in the final sprint of a race. To reach maximum speed, a cyclist will always select the highest gear and pedal as fast as possible. On the other hand, when riding up a very steep hill—for example, the notorious Alpe d'Huez climb on a mountain stage of the Tour de France—the cadence might fall substantially below 70 rpm, even if the cyclist already uses the lowest gear, but the slope of the road is too high to maintain the preferred cadence.

A racer will carefully select his or her back sprocket-wheel cluster depending on the course. A relatively flat course would necessitate a tooth range of 11 to 23 in the rear sprocket-wheel cluster; in comparison, a mountain course would require lower gears to climb up the hill and thus, for instance, a tooth range of 11 to 32 (the 11-tooth sprocket is kept for potential sprint finishes). Finally, preventing possible gear-ratio duplication will also affect the choice. In table 5.1, we saw duplication of the same gear ratio with a 14- and 19-tooth rear sprocket wheel (and chainrings with 53 and 39 teeth). This reduces the number of different gears and should be avoided. Duplicated or near-duplicated gear ratios occur on many less-expensive bicycles. Another aspect to be taken into consideration is the increments in gear inches from one gear to the next. The relative change from a lower gear to a higher gear can be expressed as a percentage. Cycling tends to feel more comfortable if nearly all gear changes have more or less the same percentage difference. For example, a change from a 13-tooth sprocket to a 15-tooth sprocket (15.4 percent) feels very similar to a change from a 20-tooth sprocket to a 23-tooth sprocket (15 percent), even though the latter has a larger absolute difference. Consistent percentage differences can be achieved if the number of teeth progress geometrically. So, you can now see how mathematics plays a significant role in optimizing our bicycle ride.

THE SPIROGRAPH TOY

The Spirograph toy is a drawing tool developed by the British engineer Denys Fisher (1918–2002). It was very popular in the late 1960s and early 1970s and is now, nearly half a century later, experiencing a revival. It won the Educational Toy of the Year Award three times in a row, from 1965 to 1967, and it was a Toy of the Year finalist again in 2014.[3] A Spirograph set basically consists of plastic rings and wheels of different sizes and, occasionally, a plastic circle template. Both the wheels as well as the rings, or the circles, in the template have gear teeth on their edges fitting together such that the wheels can roll without slippage along the circumference of other circles—outside other wheels or inside the rings or circles in the template. Each wheel or ring has several holes provided for a ballpoint pen. (See fig. 5.22.)

Figure 5.22. A version of the Spirograph toy.

Pins can be used to secure a stationary ring or wheel on an underlying piece of paper or cardboard. Another wheel is now placed such that its teeth engage with those of the pinned piece. By inserting a pen or pencil point through a hole in the movable wheel and pushing it around with its gear inside the gear of the stationary circle, you will enable the pen to draw a curve on the underlay. After a finite number of orbits of the movable wheel, the pen returns to its starting position on the underlay, and you will thus obtain a closed curve. By combining wheels and rings of different diameters and inserting the pen in holes of variable distance from the center of the wheels, you will create fantastic geometric shapes of unexpected intricacy and beauty, as can be seen in figure 5.23.

Figure 5.23. Some examples of curves produced with a Spirograph toy.

The curves traced out by the Spirograph toy are called *epitrochoids* if the rolling circle is outside the stationary circle; and they are called *hypotrochoids* if the rolling circle is inside the stationary circle. The drawn curves are special types of *roulettes*; roulettes are a family of curves that are generated by letting a curve roll without slipping on another stationary curve, and following a point that is attached to it, but not necessarily on the rolling curve. For instance, if we push a movable wheel inside or outside around a stationary one as we show it in figure 5.24, then the rolling curve is a circle of radius r and the stationary curve is a circle of radius R (if we ignore the gear teeth that are necessary to prevent slippage). Each hole in the rolling wheel represents a point whose position is fixed with respect to any point on the rolling circle. The terms *epitrochoid* and *hypotrochoid* are derived from the Greek word for "wheel," τροχός (*trochos*), with the prefix *epi* (Greek: ἐπί) meaning "over" or "upon," and the prefix *hypo* (Greek: ὑπο) standing for "under." The curve that is being drawn in figure 5.24 on the left is an epitrochoid; the one on the right is a hypotrochoid.

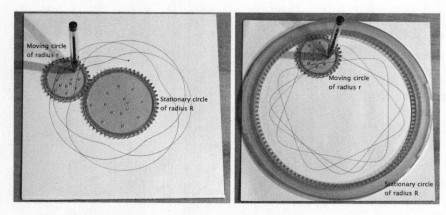

Figure 5.24. Examples of an epitrochoid *(left side)* and hypotrochoid *(right side)*, drawn with a Spirograph set.

The shape of the curve depends on the radii r and R of the two circles and on the distance d between the pen or pencil point and the center of the rotating wheel. Ignoring the absolute size of the curve, its shape (up to scaling) is actually determined by the ratios $R{:}r$ and $r{:}d$. With a Spirograph set, varying these ratios by using different combinations of rings, wheels, and holes therein, is simple and will lead to an astonishingly diverse family of curves that can be drawn. The complexity of the examples shown in figure 5.23 seems to indicate that a mathematical description of these curves must be quite complicated. However, using analytic geometry together with some basic trigonometry, you can actually derive formulas representing these curves, no matter how intertwined and complex they might appear.

In figure 5.25 we show how an epitrochoid emerges by rolling a circle of radius r without slipping on the circumference of a circle of radius R. The hole through which you insert the pen or pencil is represented by the point labeled P, at a distance d from the center C of the moving circle, which we will refer to as the "wheel."

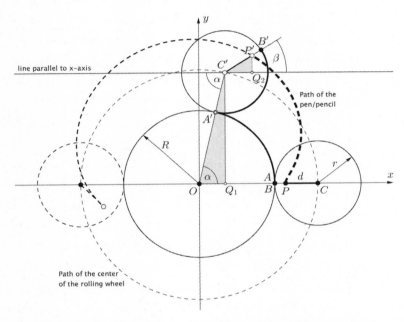

Figure 5.25. Finding the angle-dependent coordinates of a pen or pencil creating an epitrochoid.

Note that P has a fixed position on the wheel, which we indicated by a rod connecting P and C. If you let the wheel roll without slipping counterclockwise around the stationary circle, point P will trace the dashed trajectory. To calculate the path of P, you have to introduce a coordinate system. It is convenient to choose the location of the center of the stationary circle as the origin of the coordinate system. Without loss of generality, you may assume that the point of contact between both circles is initially located on the x-axis. As you roll the wheel around the stationary circle, the angle α of the center of the wheel increases and the point P traces an epitrochoid. To describe the path of P mathematically, you only have to find out how its x- and y-coordinates change with α. Figure 5.25 shows the position of the wheel at different instances in time. Point A is the point of contact between the two circles. Point B represents the point on the wheel that is initially in contact with the stationary circle and thus initially congruent with A. The primed labels refer to the positions of the respective points at a later time, when the wheel has reached an angle α. Let $x(\alpha)$ and $y(\alpha)$ represent the coordinates of P in dependence on the angle α. Clearly, the initial position of P is given by $x(0) = R + r - d$ and $y(0) = 0$. Taking a look at the right triangles $OC'Q_1$ and $C'P'Q_2$, where Q_1 and Q_2 are auxiliary points, it easy to see that $x(\alpha) = OQ_1 + C'Q_2$ and $y(\alpha) = C'Q_1 + P'Q_2$.

Now we have to invoke a little bit of trigonometry! Firstly, we notice that $OQ_1 = (R + r) \cdot \cos \alpha$ and $C'Q_1 = (R + r) \cdot \sin \alpha$. Determining the lengths of the sides of triangle $C'P'Q_2$ is a little harder. At least we can say that $C'Q_2 = d \cdot \cos\beta$ and $P'Q_2 = d \cdot \sin\beta$, where β is the angle $\angle P'C'Q_2$ (see fig. 5.25). Therefore, the task is to calculate β. Thoroughly inspecting the various angles formed around C', we observe that $\beta = \angle B'C'A' - \angle Q_2C'A'$, where $\angle Q_2C'A' = 180° - \alpha$. Thus, it only remains to find a relation between α and $\angle B'C'A'$. The key to solving this problem is to recognize that the length of the circular arc AA' must be exactly equal to the length of the circular arc $A'B'$, because the wheel rolls without slipping. Since α is the central angle of the arc AA' and $\angle B'C'A'$ is the central angle of the arc $A'B'$, we must have $R \cdot \alpha = r \cdot \angle B'C'A'$, implying $\angle B'C'A' = \frac{R}{r} \cdot \alpha$.

Therefore, we obtain $\beta = \frac{R}{r} \cdot \alpha - (180° - \alpha) = \left(\frac{R+r}{r}\right) \cdot \alpha - 180°$. Recalling our expressions for $x(\alpha)$ and $y(\alpha)$, we are actually interested in the cosine and sine of β. Employing the identities $\cos(\beta \pm 180°) = -\cos(\beta)$ and $\sin(\beta \pm 180°) = -\sin(\beta)$, we finally arrive at:

$$x(\alpha) = (R+r) \cdot \cos(\alpha) - d \cdot \cos\left(\frac{R+r}{r} \cdot \alpha\right)$$

$$y(\alpha) = (R+r) \cdot \sin(\alpha) - d \cdot \sin\left(\frac{R+r}{r} \cdot \alpha\right).$$

These formulas enable us to compute the coordinates $x(\alpha)$ and $y(\alpha)$ for any value of α, thereby obtaining a point of the corresponding epitrochoid. For hypotrochoids, where the wheel is rolling inside the stationary circle, an analogous line of arguments leads to the following:

$$x(\alpha) = (R-r) \cdot \cos(\alpha) + d \cdot \cos\left(\frac{R-r}{r} \cdot \alpha\right)$$

$$y(\alpha) = (R-r) \cdot \sin(\alpha) - d \cdot \sin\left(\frac{R-r}{r} \cdot \alpha\right).$$

Having revealed the mathematics behind the Spirograph toy, we may now investigate the generated curves more systematically. In figures 5.26, 5.27, 5.28, and 5.29, we show series of epitrochoids with fixed values for d and r, but different values for R. This means that in each of the series we always use the same wheel and the same pen hole, but we let the wheel orbit around stationary circles of different sizes. In figures 5.26, 5.27, and 5.28, the stationary circle is larger than the moving wheel ($R > r$); in figure 5.29, it is smaller ($R < r$).

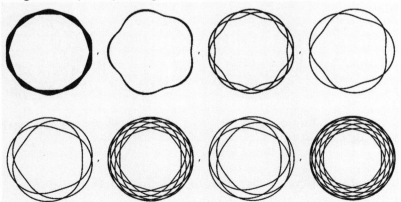

Figure 5.26. Epitrochoids for $\frac{d}{r} = \frac{2}{5}$ and $\frac{r}{R} = \frac{1}{k}$, where k is an integer between 1 and 10.

218 THE MATHEMATICS OF EVERYDAY LIFE

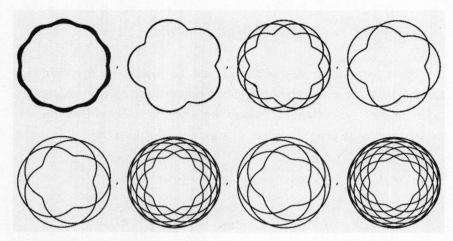

Figure 5.27. Epitrochoids for $\frac{d}{r}=\frac{2}{3}$ and $\frac{r}{R}=\frac{1}{k}$, where k is an integer between 1 and 10.

Figure 5.28. Epitrochoids for $\frac{d}{r}=\frac{17}{18}$ and $\frac{r}{R}=\frac{1}{k}$, where k is an integer between 1 and 10.

The only difference between figures 5.26, 5.27, and 5.28 is the value of parameter *d*, that is, the distance of the pen or pencil from the center of the rolling wheel. As the ratio *d*:*r* increases, the curve bends more sharply in the points closest to its center of symmetry. In figure

5.28, the ratio *d:r* is almost 1, indicating that the pen or pencil is very close the rim of the rolling wheel.

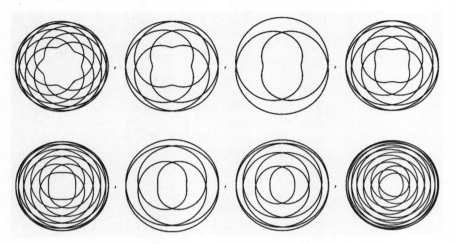

Figure 5.29. Epitrochoids for $\frac{d}{r}=\frac{3}{4}$ and the following values for $\frac{r}{R}$: $\frac{8}{7}, \frac{5}{4}, \frac{3}{2}, \frac{7}{4}, \frac{11}{4}, \frac{5}{2}, \frac{7}{2}, \frac{14}{3}$.

To complete the picture, we will look at some hypotrochoids as well. Recall that here the wheel is now rolling *inside* a stationary ring, which must therefore have a larger diameter than the moving wheel (see the picture on the right in figure 5.24).

220　THE MATHEMATICS OF EVERYDAY LIFE

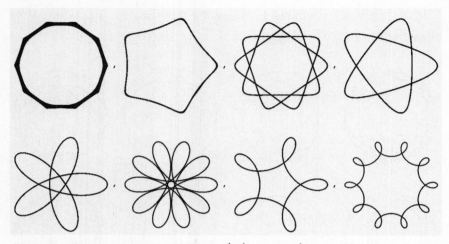

Figure 5.30. Hypotrochoids for $\frac{d}{r} = \frac{1}{2}$ and $\frac{r}{R} = \frac{1}{k}$, where k is an integer between 1 and 10.

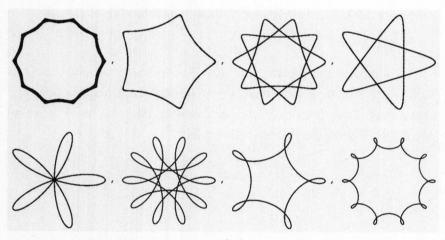

Figure 5.31. Hypotrochoids for $\frac{d}{r} = \frac{2}{3}$ and $\frac{r}{R} = \frac{1}{k}$, where k is an integer between 1 and 10.

In figure 5.31, the pen or pencil is closer to the rim of the orbiting wheel than in figure 5.30. Again, we observe that the curves in figure 5.31 change their direction more sharply than those in figure 5.30. What will happen if $d = r$? Of course, we cannot insert the pen or pencil through a hole that is exactly on the rim of the wheel, but there is no difficulty in performing this experiment on a computer. In figure 5.32 we show the resulting curves. They have sharp, thorn-like cusps, since the pen or pencil changes its direction by 180 degrees whenever it touches the stationary circle. The curves in figure 5.32 are special variants of hypotrochoids called *hypocycloids*. Analogously, an epitrochoid would become an *epicycloid*, if $d = r$.

Hypocycloids and epicycloids are derived from *cycloids*, which are curves traced by a point on the circumference of a circle as the circle rolls along a straight line (without slippage). If we increase the radius R of the stationary circle (without changing r), the ratio $r:R$ gets smaller, and as this ratio gets closer and closer to zero, our hypocycloids and epicycloids would approach cycloids. Locally, the circumference of the stationary circle will more and more look like a straight line as R gets larger and larger; we experience this ourselves simply by standing on the surface of the earth, when we cannot observe the curve from a local perspective.

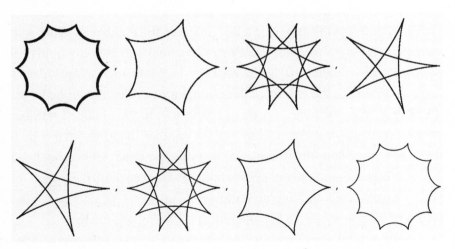

Figure 5.32. Hypotrochoids for $d = r$ and $\frac{r}{R} = \frac{1}{k}$, where k is an integer between 1 and 10.

An interesting question concerning the Spirograph toy is the following: How many times must the rolling wheel orbit around the stationary circle until the pen meets its starting point again? In other words, how many orbits do we have to draw in order to see the complete curve? The answer to this question lies in the ratio $\frac{r}{R}$, which is essentially the ratio between the numbers of gear teeth on the circumferences of the two circles. Assume that we use a stationary wheel with 150 gear teeth and a rolling wheel with 35 gear teeth revolving outside (or inside) the larger wheel. We mark the one gear tooth of the rolling circle initially in contact with the stationary wheel. After one orbit of the rolling wheel, the mark will be off its initial position by a number of gear teeth equal to the remainder of the division $\frac{150}{35}$, that is, by 10 teeth ($150 = 4 \cdot 35 + 10$). After two orbits, the mark will be off its initial position by 20 gear teeth of the rolling wheel. After three orbits, the difference will be 30 gear teeth; but after four orbits, it will only be 5 gear teeth, since the remainder of $4 \cdot \frac{150}{35}$ is 5. After how many orbits N will the mark meet its starting point again? This will be as soon as the remainder of $N \cdot \frac{150}{35}$ is equal to zero, which means that 35 is a divisor of $N \cdot 150$. The least common multiple (LCM) of 150 and 35 gives the number of gear-teeth contacts necessary to complete our epitrochoid (or hypotrochoid). In our case, this would be 1050, since $1050 = \text{LCM}(150,35)$. Now we only have to figure out the corresponding number of orbits. Since the larger wheel has 150 gear teeth, we just need to divide the obtained number of gear teeth contacts by 150 to get the number of orbits after which the pen or pencil point will return to its initial position. Hence, we need $\frac{1050}{150} = 7$ orbits to complete our epitrochoid (or hypotrochoid). Figure 5.33 shows epitrochoids and hypotrochoids for a ratio of 150:35 between the radius of stationary circle and the radius of the rolling circle.

More generally, using a smaller wheel with p gear teeth and a larger wheel or ring with q gear teeth ($q > p$), the least common multiple of p and q divided by q gives the number of orbits of the smaller wheel that are required to complete the curve, that is, $N = \frac{LCM(p,q)}{q}$.

SPORTS AND GAMES 223

Figure 5.33. Epitrochoids *(first row)* and hypotrochoids *(second row)* for a ratio of 150:35.

The diversity of shapes that can be produced with a Spirograph toy is astonishing. The surprising complexity and profound elegance that can emerge from the geometry of these curves has not only fascinated children and adults playing with a Spirograph, but also inspired engineers, architects, and artists. For younger readers, we highly recommend you get involved in a hands-on experience and use a Spirograph toy to create your own Spirograph designs. The advanced reader may try to construct epitrochoids and hypotrochoids on the computer by means of their governing equations, using geometry software such as the Geometer's Sketchpad or GeoGebra. For the less ambitious, there are also various free and ready-to-use Spirograph drawing programs available on the Internet.

Now that we have considered the mathematics around the manmade world, which largely involves recreation, we will now embark on a mathematical journey through the natural world, which we cannot control, but we ought to be able to understand better from a mathematical point of view. Hence, we continue our journey by looking at the earth from a distance and from the surface.

CHAPTER 6

THE WORLD AND ITS NATURE

The applications of mathematics as related to the world we live in are essentially boundless. In this chapter, we will show how mathematics can be used to measure, navigate, and map out the earth, and to appreciate mathematically what the earth demonstrates to us in nature. We begin with a review of the ancient method of measuring the size of the earth.

MEASURES OF AND ON THE EARTH

Have you ever stood at the beach and wondered how far you can see to the horizon line, that is, the separation between the sky and the water? Of course, this depends to some degree—although be it minimally—on the height of the observer. With simple geometry, we can calculate that distance. However, before we determine the distance to the horizon line, let's see how you could possibly measure the size of the earth, which for our purposes here, we will consider to be a perfect sphere, even though we know that there is a slight difference in the diameter measured between the North Pole and South Pole (7,898 miles) and the equatorial diameter (7,926 miles), for an average diameter of 7,912 miles. Using the formula for the circumference of a circle, πd, we find an average circumference of 24,856 miles.

226 THE MATHEMATICS OF EVERYDAY LIFE

With modern instrumentation, measuring the earth is not terribly difficult, but this was no mean feat thousands of years ago. Remember, the word *geometry* is derived from "earth measurement." Therefore, it is appropriate to consider this issue in one of its earliest forms. One of these measurements of the circumference of the earth was made by the Greek mathematician Eratosthenes (276 BCE–194 BCE) in about 230 BCE. His measurement was remarkably accurate—less than 1 percent in error![1] To make this measurement, Eratosthenes used the relationship of alternate-interior angles of parallel lines, to which we were introduced in high-school geometry.

As librarian of Alexandria, Eratosthenes had access to records of calendar events. He discovered that at noon on a certain day of the year in the town of Syene (now called Aswan) on the Nile River, the sun was directly overhead. As a result, the bottom of a deep well was entirely lit and a vertical pole, parallel to the rays hitting it, cast practically no shadow.

At the same time, however, a vertical pole in the city of Alexandria did cast a shadow. When that day arrived again, Eratosthenes measured the angle (which is shown as $\angle 1$ in figure 6.1) formed by such a pole and the ray of light from the sun going past the top of the pole to the far end of the shadow. He found it to be about $7°12'$, or $\frac{1}{50}$ of $360°$.

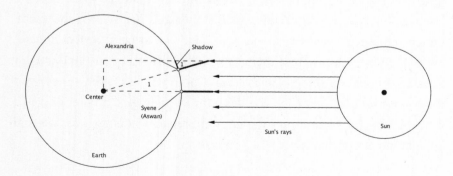

Figure 6.1.

Assuming the rays of the sun to be parallel, he knew that the angle at the center of the earth must be congruent to $\angle 1$, and, hence, must also measure approximately $\frac{1}{50}$ of 360°. Since Syene and Alexandria were almost on the same meridian or longitudinal line, Syene must be located on the radius of the circle, which was parallel to the rays of the sun. Eratosthenes thus deduced that the distance between Syene and Alexandria was $\frac{1}{50}$ of the circumference of the earth. The distance from Syene to Alexandria was believed to be about 5,000 Greek *stadia*. A *stadium* was a unit of measurement equal to the length of an Olympic or Egyptian stadium. Therefore, Eratosthenes concluded that the circumference of the earth was about 250,000 Greek stadia, or about 24,660 miles. This is very close to modern calculations, which have determined the circumference to be 24,856 miles. So how's that for some *real* geometry!

Let's consider our original problem, that of determining the distance to the horizon. Once again, we will work with the assumption that the earth is a perfect sphere. Imagine that we are standing on a beach on a perfectly clear day, ignoring any light refraction that may occur while passing through the atmosphere, and our eyes are at a height of h above the sea level, and the distance to the horizon will be designated by d, with r representing the radius of the earth.

228 THE MATHEMATICS OF EVERYDAY LIFE

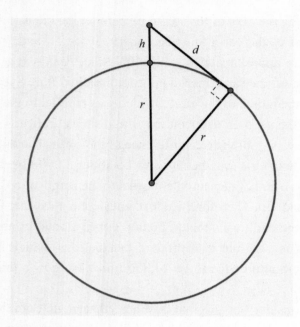

Figure 6.2.

Applying the Pythagorean theorem to the right triangle shown in figure 6.2, we get the following: $(r + h)^2 = d^2 + r^2$. Compared to the radius of the earth, our height above sea level is negligible, so that we can safely ignore h^2, without too much loss of accuracy, when we square the binomial above to get $(r + h)^2 = r^2 + h^2 + 2rh \approx r^2 + 2rh$. By then substituting this value for the binomial, we get $r^2 + 2rh = d^2 + r_i^2$, which then simplifies to $d = \sqrt{2rh}$. Let's assume that our eyes are 6 feet above the ground, which would give us $h = 6$ feet or $\frac{6}{5,280}$ miles. Therefore, with the earth's average radius of 3,956 miles, we get $d = \sqrt{(2)(3,956)\left(\frac{6}{5,280}\right)} = \sqrt{8.99} \approx 2.9$ miles for the distance from us, the viewers, to the horizon. Naturally, if we were standing on a higher platform, such as a lifeguard's seat, the distance we can see to the horizon will be considerably longer. For example, if you were to look at the horizon from the magnificent ledge where three United States presidents vacationed (Ulysses S. Grant, Chester A. Arthur, and Theodore Roosevelt) in the village of Haines Falls in the Catskill Mountains of New York State, where the famous Catskill Mountain House once stood at an elevation of 2,250 feet, according to our formula you should be able to see the horizon line at a distance of about 58 miles. To consider an extreme, suppose you were at the top of Mount Everest, at an elevation of 29,029 feet. From there, the distance to the horizon would be approximately 208 miles. You now have a method of determining whether the lyrics by Alan Jay Lerner were correct in his musical *On a Clear Day You Can See Forever*. So we can conclude that he was exaggerating a bit!

NAVIGATING THE GLOBE

There are entertainments in mathematics that stretch the mind (gently, of course) in a very pleasant and satisfying way. As we navigate the earth, we will begin with a popular puzzle question that has some very interesting extensions, and that will help us as we traverse the globe. Here we will be required to do some "out of the box" thinking, with the hope that we will leave you with some favorable impressions. Let's consider the following question: Where on Earth can you be so that you can walk 1 mile **south**, then 1 mile **east**, and then 1 mile **north** and end up at the starting point?

Perhaps after a few attempted starting points, you may come up with the right answer: the North Pole, of course! To test this answer, start from the North Pole (shown in fig. 6.3), and then travel south 1 mile, and then go east 1 mile, which takes you along a latitudinal line— one that remains equidistant from the North Pole, at a distance of 1 mile from the pole. Then, travel 1 mile north, and you will get back to where you began, the North Pole.

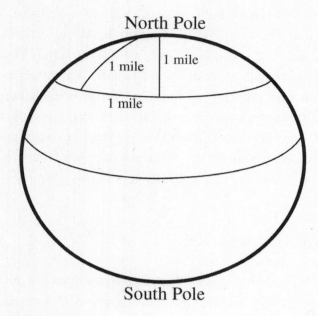

Figure 6.3.

Most people familiar with this problem feel a sense of completion. Yet we can ask, Are there other such starting points where we can take the same three directional "walks" and end up back at the starting point? The answer, surprising enough, is *yes*. One set of starting points is found by locating the latitudinal circle, which has a circumference of 1 mile, and is nearest the South Pole. (See fig. 6.4.) From this circle, walk 1 mile north (along a great circle, naturally), and form another latitudinal circle. Any point along this second latitudinal circle will qualify. Let's try it.

Begin on this second latitudinal circle (1 mile farther north). Walk 1 mile south, which takes you to the first latitudinal circle. Then walk 1 mile east, which takes you exactly once around the circle, and then 1 mile north, which takes you back to the starting point.

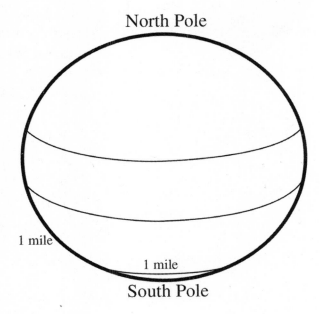

Figure 6.4.

Suppose the latitudinal circle closer to the South Pole, the one along which we would walk, would have a circumference of half a mile. We could still satisfy the given instructions, yet this time we would have to walk around the circle *twice*, and get back to our starting point on a latitudinal circle that is 1 mile farther north. If the latitudinal circle nearer to the South Pole had a circumference of a quarter of a mile, then we would merely have to walk around this circle *four* times to get back to the starting point of this circle and then go north 1 mile to the original starting point.

At this point, we can take a giant leap to a generalization that will lead us to many more points that satisfy the original stipulations. Actually, an infinite number of points! This set of points can be located by beginning with the latitudinal circle, located nearest the South Pole, that has a circumference of $\frac{1}{n}$th of a mile, so that a mile-long walk east will take you back to the point on the circle at which you began your walk on this latitudinal circle (since a mile-long walk on this circle entails walking around this circle n times). The rest is the same as before, that is, walking 1 mile south and then later 1 mile north. Is this possible with latitude circle routes near the North Pole? Yes, of course! As you can see, mathematics can be quite helpful when navigating the globe.

WHAT IS RELATIVITY?

The concept of relativity is generally not well understood by most nonscientists. Although it is often associated with Albert Einstein, it has many applications. It may be a difficult concept to grasp for some, so we shall exercise patience and support as we gently navigate further. Consider the following problem:

> ***The problem:*** **While rowing his boat upstream, David drops a cork overboard and continues rowing for 10 more minutes. He then turns around, chasing the cork, and retrieves it when the cork has traveled 1 mile downstream. What is the rate of the stream?**

Rather than approach this problem through the traditional methods that are common in an algebra course, consider the following. The problem can be made significantly easier by considering the notion of relativity.

It does not matter whether the stream is moving and carrying David downstream, or whether it is still. We are concerned only with the separation and coming together of David and the cork. If the stream were stationary, David would require as much time rowing to the cork as he did rowing away from the cork. That is, he would require 10 + 10 = 20 minutes. Since the cork travels 1 mile during these 20 minutes, the stream's rate of speed is 3 miles per hour (60 min. ÷ 20 min. = 3 and, of course, 1 mi. · 3 = 3 mi.).

Again, this may not be an easy concept to grasp for some and is best left to ponder in quiet. It is a concept worth understanding, for it has many useful applications in our thinking processes in everyday life. This is, after all, one of the purposes for understanding mathematics.

COLORING A MAP

Have you ever wondered how a map is colored? Aside from deciding which colors should be used, the question might come up regarding the number of colors that are required for a specific map so that there will never be the same color on both sides of a boarder. How many colors do you think are required? Mathematicians have determined the answer to this question: You will never need more than four different colors to color any map, regardless of how many borders or contorted arrangements the map presents. For many years, the question as to how many colors are required was a constant challenge to mathematicians, especially those doing research in topology, which is a branch of mathematics related to geometry, for which figures discussed may appear on plane surfaces or on three-dimensional surfaces. The topologist studies the properties of a figure that remain the same *after* the figure has been distorted or stretched according to a set of rules. A piece of string with its ends connected may take on the shape of a circle, or a square, which is all the same for the topologist. In going through this transformation, the order of the "points" along the string does not change. This retention of ordering has survived the distortion of shape, and it is this property that attracts the interest of topologists. Therefore, a circle and the square represent the same geometric concept to the topologist.

Throughout the nineteenth century, it was believed that five colors were required to color even the most complicated-looking map. However, there was always strong speculation that four colors would suffice. It was not until 1976 that the mathematicians Kenneth Appel (1932–2013) and Wolfgang Haken (1928–) proved that four colors were sufficient to color any map.[2] However, unconventionally, they used a high-powered computer to consider all possible map arrangements. It must be said that there are still mathematicians who are dissatisfied with their "proof," since it was done by computer and not in the traditional way, "by hand." Previously, it was considered one of the famous unsolved problems of mathematics. Let us now delve into the consideration of various maps and the number of colors required to color them in such a way that no common boundary of two regions shares the same color on both of its sides.

Suppose we consider a geographic map that has a configuration analogous to that shown in figure 6.5.

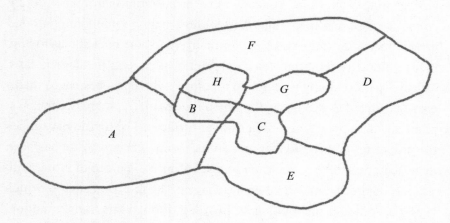

Figure 6.5.

Here we notice that there are eight different regions indicated by the letters shown. Suppose we list all regions that have a common boundary with region H, and regions that share a common vertex with region H. The regions designated by the letters B, G, and F share a border with region H. The region designated by the letter C shares a vertex with the region designated by the letter H.

THE WORLD AND ITS NATURE   235

Remember, a map will be considered correctly colored when each region is completely colored and two regions that share a common boundary have different colors. Two regions sharing a common vertex may share the same color. Let's consider coloring a few maps (see fig. 6.6) to see various potential map configurations that require no more than three colors. (note: **b** indicates blue; **r** indicates red; **y** indicates yellow; and **g** indicates green).

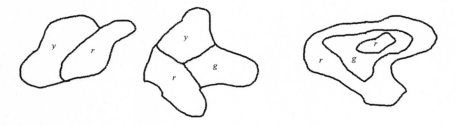

Figure 6.6.

The first map in figure 6.6 is able to be colored in two colors: yellow and red. The second map required three colors: yellow, red, and green. The third map has three separate regions but requires only two colors, red and green, since the innermost territory does not share a common border with the outermost territory. It would seem reasonable to conclude that if a three-region map can be colored with fewer than three colors, a four-region map can be colored with fewer than four colors. Let's consider such a map.

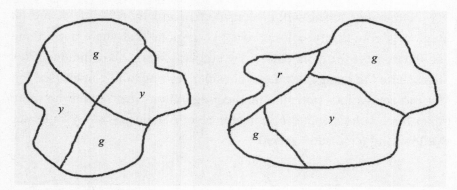

Figure 6.7.

The left-side map shown in figure 6.7 has four regions and requires only two colors for correct coloring, whereas the right-side map also consists of four regions but requires three colors for correct coloring.

We should now consider a map that requires four colors for proper coloring of the regions. Essentially, this will be a map in which each of the four regions shares a common border with the other three regions. One possible such mapping is shown in figure 6.8.

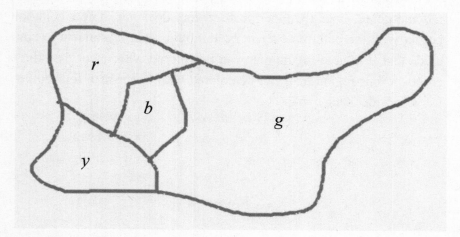

Figure 6.8.

If we now take the next logical step in this series of map-coloring challenges, we should come up with the idea of coloring maps involving five distinct regions. It will be possible to draw maps that have five regions and require two, three, or four colors to be colored correctly. The task of drawing a five-region map that *requires* five colors for correct coloring will be impossible. This curiosity can be generalized through further investigation and should convince you of the idea that any map on a plane surface, with any number of regions, can be successfully colored with four or fewer colors. You might want to challenge friends to conceive of a map of any number of regions that requires more than four colors so that no two regions with a common border share the same color.

This puzzle has remained alive for many years, challenging some of the most brilliant minds; but, as we said earlier, the issue has been closed by the work of the two mathematicians Appel and Haken. There are still many conjectures in mathematics that have escaped proof but have never been disproven. Here, at least, we have a conjecture that is closed.

CROSSING BRIDGES

New York City is comprised mostly of islands—as a matter of fact, the only part of the city that is part of the mainland of the United States is the Bronx, excluding City Island, of course. Consequently, the city is blessed with many bridges. There are bike and track races that traverse several bridges throughout their paths. Most of us take bridge crossing for granted these days. Bridges essentially become part of the path traveled and escape our attention until there is a toll to be paid. Then you take particular note of the bridge crossing, of course.

In the eighteenth century and earlier, when walking was the dominant form of local transportation, people would often count particular kinds of objects they passed. One such was bridges. Through the eighteenth century, the small Prussian city of Königsberg (today called Kaliningrad, Russia), located where the Pregel River forms two branches around an island portion of the city, provided a recreational dilemma: Could a person walk over each of the seven bridges exactly once in a continuous walk through the city? The residents of the city had this as an entertaining activity, particularly on Sunday afternoons. Since there were no successful attempts, the challenge continued for many years.

This problem provides a wonderful window into networks, which is referred to as *graph theory*, an extended field of geometry that gives us a renewed view of the subject. To begin, we should present the problem. In figure 6.9, we can see a map of the city with the seven bridges.

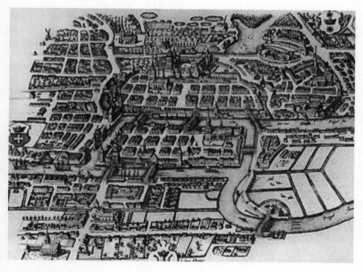

Figure 6.9.

In figure 6.10, we will indicate the island by Ⓐ, the left bank of the river by Ⓑ, the right bank by Ⓒ, and the area between the two arms of the upper course by Ⓓ. The seven bridges are called Holz, Schmiede, Honig, Hohe, Köttel, Grüne, and Krämer. If we start at Holz and walk to Schmiede and then through Honig, through Hohe, through Köttel, and then through Grüne, we will never cross Krämer. On the other hand, if we start at Krämer and walk to Honig, through Hohe, through Köttel, through Schmiede, and then through Holz, we will never travel through Grüne.

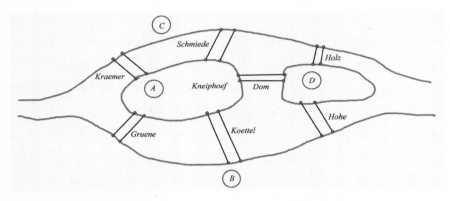

Figure 6.10.

In 1735, the famous mathematician Leonhard Euler (1707–1783) proved mathematically that this walk could not be performed.[3] The famous Königsberg Bridges Problem, as it became known, is a lovely application of a topological problem with networks. It is very nice to observe how mathematics used properly can put a practical problem to rest. Before we embark on the problem, we ought to become familiar with the basic concept involved. Toward that end, try to trace with a pencil each of the following configurations without missing any part and without going over any part twice. Keep count of the number of arcs or line segments, which have an endpoint at each of the points A, B, C, D, and E.

Configurations such as the five figures, called networks, which are shown in figure 6.11, are made up of line segments and/or continuous arcs. The degree of the vertex is determined by the number of arcs or line segments that have an endpoint at that particular vertex, and, as such, the degree of the vertex can be odd or even.

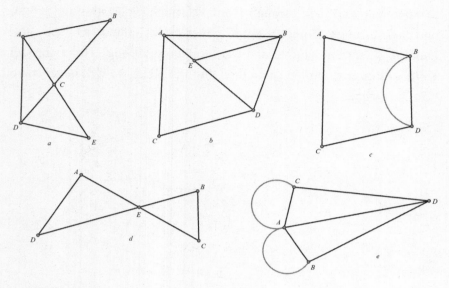

Figure 6.11.

After trying to trace these networks without taking the pencil off the paper and without going over any line more than once, you should notice two direct outcomes. The networks can be traced (or traversed) if they have (1) all even-degree vertices or (2) exactly two odd-degree vertices. The following two statements summarizes this:

1. There is an even number of odd-degree vertices in a connected network.
2. A connected network can be traversed only if it has at most two odd-degree vertices.

Network figure 6.11a has five vertices. Vertices B, C, and E are of even degree, and vertices A and D are of odd degree. Since figure 6.11a has exactly two odd-degree vertices as well as three even-degree vertices, it is traversable. If we start at A then go down to D, across to E, back up to A, across to B, and down to D, we have chosen a desired route.

Network figure 6.11b has five vertices. Vertex C is the only even-degree vertex. Vertices A, B, E, and D are all of odd-degree. Consequently, since the network has more than two odd-degree vertices, it is not traversable.

Network figure 6.11c is traversable because it has two even-degree vertices and exactly two odd-degree vertices.

Network figure 6.11d has five even-degree vertices and, therefore, can be traversed.

Network figure 6.11e has four odd-degree vertices and cannot be traversed.

The Königsberg Bridges Problem is the same problem as the one posed in figure 6.11e. Let's take a look at figure 6.11e and figure 6.10 and note the similarity. There are seven bridges in figure 6.10, and there are seven lines in figure 6.11e. In figure 6.11e, each vertex is of odd-degree. In figure 6.10, if we start at Ⓓ we have three choices, we could go to Hohe, Honig, or Holz. If in figure 6.11e we start at D we have three line paths to choose from. In both figures, if we are at C or Ⓒ, we have three bridges we could go on (or three lines). A similar situation exists for locations Ⓐ and Ⓑ in figure 6.10 and vertices A and B in figure 6.11e. We can see that this network cannot be traversed.

By reducing the bridges and islands to a network problem, we can easily solve it. This is a clever tactic to solve problems in mathematics. You might try to find a group of local bridges to create a similar challenge and see if the walk is traversable. This problem and its network application is an excellent introduction into the field of topology, and, beyond that, we can also apply this technique of the traversibility of a network to the famous "Five-Bedroom-House Problem." Let's consider the floor plan of a house with five rooms as shown in figure 6.12.

242 THE MATHEMATICS OF EVERYDAY LIFE

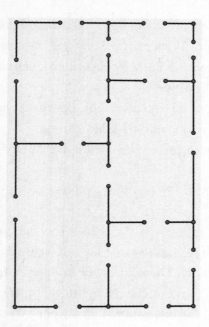

Figure 6.12.

Each room has a doorway to each adjacent room and at least one doorway leading outside the house. The problem is to have a person start either inside or outside the house and walk through each doorway exactly once. You will realize that, although the number of attempts is finite, there are far too many potential paths to make a trial-and-error solution practical.

Figure 6.13 shows various possible paths joining the five rooms A, B, C, D, and E and the outside area F.

THE WORLD AND ITS NATURE   243

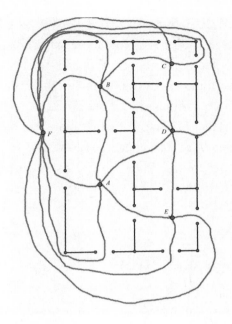

Figure 6.13.

As before, our question can be answered by merely determining whether this network is traversible. In figure 6.13 we notice that we have four odd-degree vertices, and two even-degree vertices. Since there are not exactly two or zero vertices of odd order, this network cannot be traversed. Therefore, the Five-Bedroom-House Problem does not have a solution path that would allow walking through each doorway exactly once. As you can see, even in choosing paths for travel, mathematics seems to provide a solution to our question.

MATHEMATICS IN NATURE

In this section, we will take you through a variety of unexpected occurrences of mathematics showing itself in nature. Primarily the mathematics that applies here is based on the famous Fibonacci numbers.

These numbers (1, 1, 2, 3, 5, 8, 13, 21, 34, 55, 89, 144, 233, . . .) originate from a problem about the regeneration of rabbits posed by Leonardo of Pisa (known today as Fibonacci) in chapter 12 of his book titled *Liber Abaci*. These numbers continue indefinitely and are generated by adding the last two numbers in the sequence to get the next one. In other words, the number following 233 is found by adding 144 + 233 = 377. These numbers seem to appear everywhere in nature, as you will see in the next several examples. For more about the Fibonacci numbers, see *The Fabulous Fibonacci Numbers*, by A. S. Posamentier and I. Lehmann.[4]

THE MALE BEE'S FAMILY TREE

Of the more than thirty thousand species of bees, the most well-known is probably the honeybee, which lives in a bee hive and has a family. So, let us take a closer look at this type of bee. Curiously enough, an inspection of the family tree of the male bee will reveal our famous numbers. To examine the family tree closely, we first must understand the peculiarities of the male bee. In a bee hive, there are three types of bees: the male bees (called drones) who does no work, the female bees (called worker bees) who do all the work, and the queen bee, who produces eggs to generate more bees. Male bees hatch from an unfertilized eggs, which means they have only a mother and no father (but do have a grandfather), whereas female bees hatch from fertilized eggs, thus requiring a mother (queen bee) and a father (one of the drones). The female bees end up as worker bees, unless they have taken some of the "royal jelly" that enables them to become a queen bee and begin a new colony of bees elsewhere.

In figure 6.14, where ♂ represents the male and ♀ represents the female, we begin with the male bee whose ancestors we are tracking. As we said, the male bee must come from an unfertilized egg, so only

THE WORLD AND ITS NATURE 245

a female was necessary to produce him; we see the male bee at the top and his mother in the second row. In contrast to the male bee, the egg-producing female that was its mother must have had a mother and father, so the third line has both a male and a female. It then continues with this pattern: the immediate ancestor of every male is a single female, and the immediate ancestors of a female are a male and a female. At the right of the figure is a summary of the number of bees in each row. As you look down in each column at the right, you will recognize the familiar Fibonacci numbers.

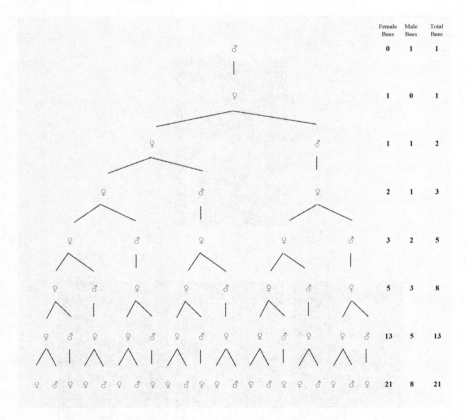

Figure 6.14.

FIBONACCI NUMBERS IN THE PLANT WORLD

Perhaps you would be most surprised to discover that these ubiquitous numbers also appear in the plant world. Take for example the pineapple. It might even be worth your while to get a pineapple and see for yourself that the Fibonacci numbers are represented on it. The hexagonal bracts on a pineapple can be seen to form three different-direction spirals. In figures 6.15, 6.16, 6.17, and 6.18 you will notice that in the three directions there are 5, 8, and 13 spirals. These are three consecutive Fibonacci numbers.

Figure 6.15.

Figure 6.16.   Figure 6.17.   Figure 6.18.

THE PINE CONE AND OTHERS

There are various species of pine cones (e.g., Norway spruce; Douglas fir or spruce; and Larch). Most pine cones will have two distinct-direction spirals. Spiral arrangements are classified by the number of visible spirals (parastichies) that they exhibit. The number of spirals in each direction will be most often two successive Fibonacci numbers. The two pictures of figure 6.19 bear this out, but you may want to convince yourself by finding some actual pine cones and counting the spirals yourself.

Figure 6.19.

These pine cones have 8 spirals in one direction and 13 in the other direction. Again, you will notice that these are Fibonacci numbers. Table 6.1 shows how the Fibonacci numbers can be found on pine cones of other species of pine trees.

Tree (Species)	Number of Spirals in Right Direction	Number of Spirals in Left Direction
Norway spruce	13	8
Douglas fir or spruce	3	5
Larch	5	3
Pine	5	8

Table 6.1.

The following designations of spiral patterns do not make any pretense of completeness. Actually, bracts can be lined up into sequences in many ways. The following are simply some of the more-obvious patterns that have been observed. (Regarding the notation, 8-5, for example, means that starting from a given bract and proceeding along two different-direction spirals, 8 bracts will be found for one spiral and 5 for the other.)

Pinus albicaulis (whitebark pine)	5-3, 8-3, 8-5
Pinus flexilis (limber pine)	8-5, 5-3, 8-3
Pinus lambertiana (sugar pine)	8-5, 13-5, 13-8, 3-5, 3-8, 3-13, 3-21
Pinus monticola (Western white pine, silver pine)	3-5
Pinus monophylla (single-leaf pinyon)	3-5, 3-8
Pinus edulis	5-3
Pinus quadrifolia (four-leaf pinyon)	5-3
Pinus aristata (bristlecone pine)	8-5, 5-3, 8-3
Pinus Balfouriana (foxtail pine)	8-5, 5-3, 8-3
Pinus muricata (bishop pine)	8-13, 5-8
Pinus remorata (Santa Cruz Island pine)	5-8
Pinus contorta (shore pine)	8-13
Pinus murrayana (lodgepole pine, tamarack pine)	8-5, 13-5, 13-8

Pinus torreyana (Torrey pine)   8-5, 13-5
Pinus ponderosa (Yellow pine)   13-8, 13-5, 8-5
Pinus jeffreyi (Jeffrey's pine)   13-5, 13-8, 5-8
Pinus radiata (Monterey pine)   13-8, 8-5, 13-5
Pinus attenuata (knobcone pine)   8-5, 13-5, 3-5, 3-8
Pinus sabiniana (digger pine)   13-8
Pinus coulteri (Coulter pine)   13-8

We can find spiral patterns, especially 5-8, in Aroids. The aroids (family Araceae) are a group of attractive ornamental plants that include the very familiar *Aglaonemas*, *Alocasias*, *Anthuriums*, *Arums*, *Caladiums*, *Colocasias*, *Dieffenbachias*, *Monsteras*, *Philodendrons*, *Scindapsuses*, and *Spathiphyllums*.[5] Spirals are also found on a variety of other plants. Figures 6.20, 6.21, 6.22, 6.23, and 6.24 are a few representative examples. You may well want to search for others.

Figure 6.20. Mammillaria Huitzilopochtli has 13 and 21 spirals.

Figure 6.21. Mammillaria Magnimamma has 8 and 13 spirals.

250 THE MATHEMATICS OF EVERYDAY LIFE

Figure 6.22. Marguerite has 21 and 34 spirals.

Figure 6.23. Sunflower has 34 and 55 spirals.

Figure 6.24. Sunflower has 55 and 89 spirals.

Based on a survey of the literature encompassing 650 species and 12,500 specimens, Roger V. Jean[6] estimated that, among plants displaying spiral, or multijugate phyllotaxis, about 92 percent of them have Fibonacci phyllotaxis.

You may wonder why this happens. Certainly, mathematicians

did not influence this phenomenon. The reason seems to be that this arrangement forms an optimal packing of the seeds so that, no matter how large the seed range, if all of the seeds being the same size, they should be uniformly packed at any stage (this means that there is no crowding in the center and the seeds are not too sparse near the edges). We "see" this packing as spirals, which almost always translates into adjacent Fibonacci numbers.

LEAF ARRANGEMENT—PHYLLOTAXIS

Until now, we have concentrated on the center of the sunflower, or, for that matter, the daisy; but we can now inspect the petals surrounding the center. Again, you will discover that most plants will have a number of petals corresponding to a Fibonacci number. For example, lilies and irises have 3 petals; buttercups have 5 petals; some delphiniums have 8; corn marigolds have 13 petals; some asters have 21 petals; and daisies can be found with 34, 55, or 89 petals—all Fibonacci numbers!

Here is a short list of some flowers arranged by the number of petals they have:

3 petals: iris, snowdrop, lily (some lilies have 6 petals formed from two sets of 3)
5 petals: buttercup, columbine (*Aquilegia*), wild rose, larkspur, pinks, apple blossom, hibiscus
8 petals: delphiniums, *Cosmos bipinnatus*[7], *Coreopsis tinctoria*
13 petals: corn marigold, cineraria, some daisies, ragwort
21 petals: aster, black-eyed Susan, chicory, *Helianthus annuus*
34 petals: plantain, pyrethrum, some daisies
55, 89 petals: Michaelmas daisies, the Asteraceae family

Some species, such as the buttercup, are very precise about the number of petals they have, but other species have petals whose numbers are very near those above, with the average being a Fibonacci number!

Having now inspected the parts of flowers, we can look at the placement of the leaves on a stem. Take a plant that has not been pruned, and locate the lowest leaf. Begin with the bottom leaf, and count the number of rotations around the stem, each time going through the next leaf up the stem, until you reach the next leaf whose direction is the same as the first leaf you identified (that is, above it and pointing in the same direction). The number of rotations will be a Fibonacci number. Furthermore, the number of leaves that you will pass along the way to reach the "final" leaf will also be a Fibonacci number.

For example, in figure 6.25 it took 5 revolutions to reach the leaf (the 8th leaf) that is in the same direction as the first one. This phyllotaxis (i.e., leaf arrangement) will vary with different species but should be a Fibonacci number. If we refer to the rotation/leaf-number ratio for this plant, it would be 5/8. The curve marked in figure 6.25 is described as the "genetic spiral of a plant."

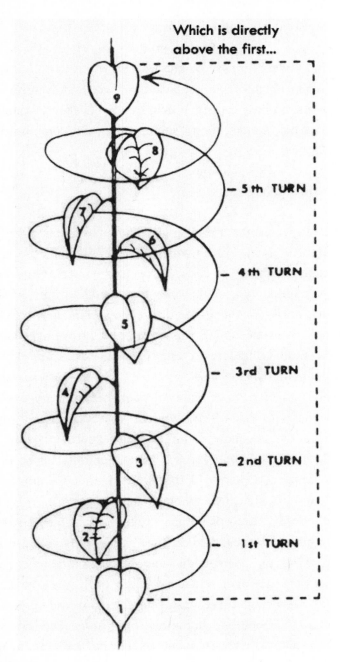

Figure 6.25.

Here are some phyllotaxis ratios:

1/2: elm, linden, lime, some grasses
3/8: asters, cabbages, poplar, pear, hawkweed, some roses
1/3: alders, birches, sedges, beech, hazel, blackberry, some grasses
2/5: roses, oak, apricot, cherry, apple, holly, plum, common groundsel
8/21: fir trees, spruce
5/13: pussy willow, almond
13/34: some pine trees

Different species of palms display different numbers of leaf spirals, but the numbers always match with Fibonacci numbers. For example, in the areca nut palm (*Areca catechu*), or the ornamental *Ptychosperma macarthurii* palm, only a single foliar spiral is discernible, while in the sugar palm (*Arenga saccharifera*), or *Arenga pinnata*, 2 spirals each are visible. In the Palmyra palm (*Borassus flabellifer*), or *Corypha elata*, as well as in a number of other species of palms, 3 clear spirals are visible. The coconut palm (*Cocos nucifera*) as well as *Copernicia macroglossa* have 5 spirals, while the African oil palm (*Elaeis guineensis*) bears 8 spirals. The wild date palm (*Phoenix sylvestris*) and a few other species of palms also show 8 spirals. On stout trunks of the Canary Island palm (*Phoenix canariensis*), thirteen spirals can be observed. Also, in some of these plants, 21 spirals can be found. Palms bearing 4, 6, 7, 9, 10, or 12 obvious leaf spirals are not known.[8]

This phyllotaxis ratio of Fibonacci numbers is not guaranteed for every plant but can be observed on most plants. Why do these arrangements occur? We can speculate that some of the cases may be related to maximizing the space available to each leaf, or the average amount of light falling on each one. Even a tiny advantage would come to dominate, over many generations. In the case of close-packed leaves in cabbages and succulents, such arrangements may be crucial for availability of space.

THE WORLD AND ITS NATURE

The ubiquitous appearances of the Fibonacci numbers in nature are just another piece of evidence that these numbers have truly phenomenal qualities. These numbers show how mathematics is all around us in nature.

THE FIBONACCI NUMBERS ON THE HUMAN BODY

Even the human body exhibits mathematics via the Fibonacci numbers. For example, a typical human being has 1 head, 2 arms, and 3 joints on each finger, and 5 fingers on each arm—these are all Fibonacci numbers. In all fairness, we must take notice that the "small" Fibonacci-numbers are just pure coincidence, because among eight numbers from 1 to 8 there are five Fibonacci numbers. Therefore, you have a good chance to find a Fibonacci number purely by coincidence. The percentage of Fibonacci numbers among the rest of all other numbers changes dramatically if we look at bigger intervals of numbers.

At the end of the nineteenth century, many scientists were of the opinion that the golden ratio, ϕ—which can be obtained by taking the (limiting) ratio of consecutive Fibonacci numbers—was a divine, universal law of nature. So, how then, they wondered, can the human being, the "pride of creation," not have been designed by the rules of the golden section? Some saw it as if it were; for instance, Leonardo da Vinci (1452–1519) constructed the proportion of the human body based on the golden section, ϕ (see the Vitruvian Man, figs. 7.7 and 8.32). Furthermore, he partitioned the face vertically into thirds: the forehead, middle face, and chin. Plastic surgeons further divide the face horizontally in fifths. In this way, we see once again how the presence of the two Fibonacci numbers 3 and 5 come into play.

Da Vinci wasn't the only person who found the Fibonacci numbers in the human form. The Belgian mathematician and astronomer and founder of the social statistics, Lambert Adolphe Jacques Quetelet[9]

(1796–1874), and the German author and philosopher Adolf Zeising[10] (1810–1876) measured the human body and saw its proportions related to the golden ratio, and thus greatly influenced future generations.

The French architect Le Corbusier (Charles-Édouard Jeanneret-Gris, 1887–1965) assumed that human body proportions are based on the golden ratio. Le Corbusier formulated the ideal proportions in the following way:

height: 182 cm;
navel height: 113cm;
fingertips with arms upraised: 226 cm.

The ratio of height to navel height is $\frac{182}{113} \approx 1.610619469$, a very close approximation of ϕ. By another measurement of the same features, he found an analogous ratio of $\frac{176\,cm}{109\,cm} \approx 1.6147$, which is another good approximation of the ratio of the two Fibonacci numbers 13 and 21, or $\frac{21}{13} \approx 1.615384615$.

In 1948, Le Corbusier published a book titled *The Modulor: A Harmonious Measure to the Human Scale Universally Applicable to Architecture and Mechanics*, which caused somewhat of a furor in the art world. Le Corbusier developed a technique based on the golden ratio, in which he selected the height of a door to have a measure of 2.26 m so that a person of height 1.83 m can touch the top with outstretched arms. Le Corbusier supported his ideas with the following: "A man with a raised arm provides the main points of space displacement—the foot, solar plexus, head, and fingertip of the raised arm—in three intervals which yield a number of golden ratios that are determined by the Fibonacci numbers."[11]

The *Modulor*, which you may view online, consists of two scales, or bands, which are marked with two sequences of natural numbers:

the sequence: (*) 6, 9, 15, 24, 39, 63, 102, 165, 267, 432, 698, 1130, 1829,

and an intermediate sequence: (**) 11, 18, 30, 48, 78, 126, 204, 330, 534, 863, 1397, 2260.

At this point you may be wondering, How do these numbers in any way reflect the golden ratio? Here the author provides us with a little trick. That is, if we divide the members of the sequence (*) by 3, we get 2, 3, 5, 8, 13, 21, 34, 55, 89, 144, $232\frac{2}{3}$, $376\frac{2}{3}$, $609\frac{2}{3}$. Then, if we divide the members of the sequence (**) by 6, we get $1.8\overline{3}$, 3, 5, 8, 13, 21, 34, 55, 89, $143.8\overline{3}$, $232.8\overline{3}$, $376.8\overline{3}$. By rounding off the fractional members of each sequence, we find that we are once again dealing exclusively with the Fibonacci numbers.

This relationship to the Fibonacci numbers results from the fact that Le Corbusier's sequences almost approximate a geometric progression with the common ratio being the golden ratio. With his *Modulor*, Le Corbusier claims to have found a scale that adapts exceptionally well to the visual arts.

We can attach all sorts of numbers to parts of the human body. For example, we know that there are 32 teeth and 206 bones in the human body, and then it is estimated that there are 60,000 miles of blood vessels in the human body. There are also estimates for the number of hairs on the head of the human body: for blonde hair, the estimate is about 150,000; for brown hair, the estimate is 110,000; for black hair the estimate is 100,000, and for red hair the estimate is about 90,000. Additionally, it is estimated that there are 100 trillion bacteria living inside and on each human body.

We could go on endlessly finding numerical relationships related to the human body, but we leave it to the reader to discover more.

THE GEOMETRY OF RAINBOWS

Have you ever tried to get to the end of a rainbow? A common legend in Irish folklore asserts that a pot of gold is to be found at the end of a rainbow, but it is guarded by a leprechaun. Unfortunately, it is absolutely impossible to reach the end of a rainbow, despite the fact that you will find an astonishing number of people on the Internet who have claimed to have found the end of a rainbow—not to mention those who swear to have seen a leprechaun there. To understand why it is impossible to get to the end of a rainbow, you must first understand how, and under which circumstances, a rainbow is created. Starting with a little input from physics, we can use geometry to explain how this intriguingly beautiful phenomenon arises. In fact, a geometrical description of the formation of a rainbow, capturing all the main aspects, is quite simple. Additionally, invoking a few basic concepts from differential calculus will reveal one or two remarkable facts about rainbows you may not have been aware of.

Let us start with some basic facts with which you may already be familiar. When you are looking at a rainbow, you are actually looking at light that is reflected by water droplets. This rather-imprecise statement already implies two requirements for a rainbow to be formed: We need a strong light source and we need water droplets in the air. It is then natural to conclude that the appearance of a rainbow will depend on the position of the light source and the position of the water droplets. In the following, we will consider rainbows caused by sunlight and rain, but our discussion applies just as well for rainbows caused by other light sources and other water sources. Since the sunlight is reflected back from the water droplets, when we see a rainbow, the sun must always be at our back. As trivial as this observation might seem, it already indicates that, in addition to the position of the sun and the rain, there is a third variable that plays an equally important role, namely, the position of the observer. Later, we will later see that this is indeed a decisive factor in determining the possible position and size of a rainbow. In fact, two observers at different positions will never see the same rainbow. Quite literally, each observer sees his or her own rainbow. What this essentially means will become clear in the mathematical analysis of the problem. Not being aware of this interdependence leads to the erroneous belief that you could, at least in principle, walk to the end of a rainbow.

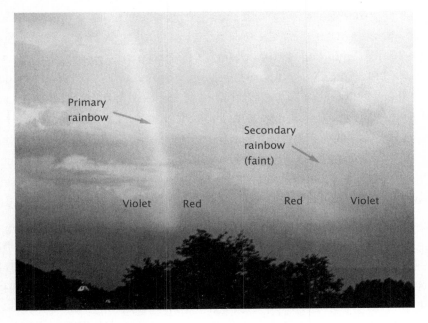

Figure 6.26.

Sometimes a second, fainter rainbow appears above the main one. Figure 6.26 shows a photograph of a "double rainbow," albeit in black and white. If you have ever seen a double rainbow in nature, you may have noticed that the colors of the secondary rainbow are reversed compared to the primary rainbow. In the primary rainbow, the inner arc is violet and the outer arc is red, while the secondary rainbow has a red inner arc and a violet outer arc. Our geometrical analysis will also explain the occurrence of secondary rainbows. The framework we will use as the basis for our investigation is a branch of physics called *geometrical optics*. In geometrical optics, the propagation of light is described in terms of rays, with certain simplifying assumptions for the interaction of light with matter. Since light is an electromagnetic wave, geometrical optics cannot explain optical phenomena for which the wave nature of light is relevant, such as, for example, interference of light waves; however, it is a very good approximation whenever the size of structures with which the light interacts is large compared to the wavelength of the light. The wavelength of light that is visible for the human eye ranges from about 390 nanometers for violet light to about 700 nanometers for red light (1 nanometer is 10^{-9} meters, that is, 0.000000001 meters).[12] The size of raindrops, on the other hand, varies from about 0.5 mm to at most 5 mm, hence the diameter of a raindrop is at least 1,000 times larger than the wavelength of visible light (we encourage you to verify this rough estimate by a short calculation).[13] It is, therefore, fine to apply the methods of geometrical optics for the description of optical effects that may arise when sunlight hits a raindrop.

Within the geometrical framework, we will make further simplifying assumptions. First, we consider raindrops as perfect spheres (although raindrops start out high in the atmosphere as approximately spherical, they lose their rounded shape as they fall). Second, we will assume that the rays of sunlight hitting a rain shower are all parallel to each other. We can justify this assumption with the help of another rough estimate that may be quite instructive, so we will digress to exhibit it. The diameter of the sun, s, is approximately 1.4 million kilometers, and the sun's average distance to the earth, d, is 150 million kilometers. The diameter of the earth, e, is about 13,000 kilometers.[14] Figure 6.27 shows a schematic diagram from which we can easily calculate the maximum angle between any two light rays emitted by the sun and hitting the earth (in physics, this is called the *divergence* of the light rays).

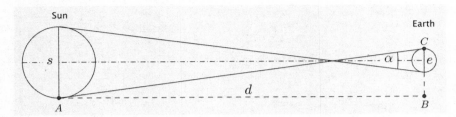

Figure 6.27.

To find the angle α, we consider the right triangle ABC. We have $\angle ABC = \frac{\alpha}{2}$ and $\sin\frac{\alpha}{2} = \frac{s+e}{2d}$. Solving for α and substituting the values for s, e, and d, we get $\alpha \approx 0.5°$, which is so small that for our purposes we can safely regard the sunlight that hits the earth as being composed of parallel rays. There is one more ingredient we need before we can delve into the geometry of rainbows. When a ray of light strikes a surface of water, a portion of it is reflected at the surface, and a portion is refracted (or "bent") into the water. You can see the refraction of light at the boundary surface between air and water, when you put a drinking straw in a glass of water, or when you stand upright in a shallow swimming pool and look at your legs. The physical reason for the refraction of light as it passes through a boundary between two different media is that the velocity of light is different in different media. If c is the speed of light in vacuum, then $\frac{c}{n}$ is the speed of light in a medium with refractive index n. The precise relationship between the angle of incidence and the angle of refraction is described by Snell's law, named after Dutch astronomer Willebrord Snellius (1580–1626). The angles are measured with respect to the normal (or perpendicular) of the tangent plane at the point where the ray strikes the surface.

262   THE MATHEMATICS OF EVERYDAY LIFE

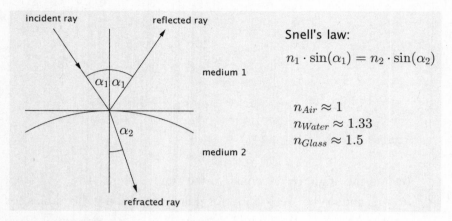

Figure 6.28.

If a ray of light passes from air (medium 1) to water (medium 2), then the angle of refraction α_2 can be expressed in terms of the angle of incidence α_1 as $\alpha_2 = \arcsin\left(\frac{\sin \alpha_1}{n}\right)$, where $n = 1.33$. Note that for an air-water boundary, the angle of refraction will always be smaller than the angle of incidence since $n > 1$, that is, the ray will be refracted toward the normal. We may also reverse the direction of the arrows in figure 6.28; that is, a light ray passing from water to air will be refracted away from the normal. A portion of the light will be reflected at the boundary, where the angle of reflection is equal to the angle of incidence (by the law of reflection).

We are now prepared to consider a light ray hitting a spherical raindrop. Figure 6.29 depicts this situation in the plane that is defined by the light ray and the center of the raindrop. An incident ray of light hits the drop of water at point A at an angle α, is refracted into the drop at an angle $\beta = \arcsin\left(\frac{\sin \alpha}{n}\right)$, and then proceeds to point B. Of course, some of the incident light rays are reflected at A, but they are not relevant for our considerations. At point B, the refracted ray is reflected to point A', where it is finally refracted back into the air. Again, some rays will already be refracted into the air at point B, and some will be reflected back into the drop at point A', but those cases are of no relevance for the primary rainbow, so we have omitted them in the diagram.

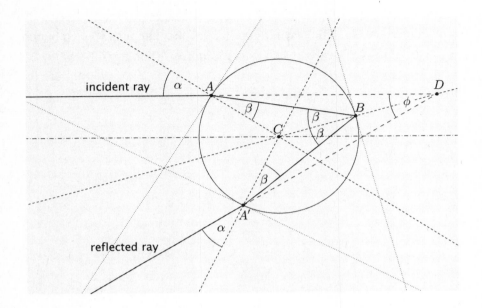

Figure 6.29.

Let us consider how the angles at points B and A' can be determined: Since $AC = CB$, which makes ABC an isosceles triangle, the angle of incidence at point B is β. Here the ray is reflected with the same angle (by the law of reflection). Then, since $A'BC$ is also an isosceles triangle, the angle of incidence at point A' must be β as well. Finally, as Snell's law works in both directions, the angle of refraction for the ray leaving the drop at point A' must be equal to α again. To find the angles at A', we could also argue that the path of the ray must be symmetric with respect to line CB. What we are actually interested in is the angle ϕ, measured between the incident ray and the reflected ray as labeled in figure 6.29. Using triangle ACD, we can write $\frac{\phi}{2} = 180° - \alpha - \angle ACB$; and, from triangle ACB, we get $\angle ACB = 180° - 2\beta$. Combining these two equations, we obtain $\phi = 2 \cdot (2\beta - \alpha)$. Upon the substitution $\beta = \arcsin\left(\frac{\sin\alpha}{n}\right)$, this becomes $\phi = 4 \cdot \arcsin\left(\frac{\sin\alpha}{n}\right) - 2\alpha$, which represents the angle between the ray striking the raindrop at A and the ray leaving the raindrop at point A' after two refractions and one reflection. Note that we have only considered one specific ray in figure 6.29. What happens when a bunch of parallel rays of sunlight strikes the raindrop? (See fig. 6.30.)

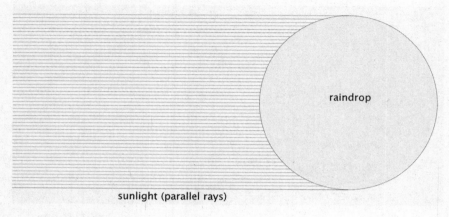

Figure 6.30.

Each ray has a different angle of incidence and will, therefore, be refracted at a different angle. Recall, at each boundary interaction (labeled by A, B, and A' in fig. 6.29) only a portion of the light will follow the path we have drawn. Some light is already reflected at point A, some is refracted back into the air at point B, and some is reflected back into the drop at point A'. Thus, the total intensity of light leaving the drop after exactly two refractions and one reflection will be significantly lower than the total intensity of the incident light. Nevertheless, it is exactly this portion of the rays that forms the primary rainbow. But why can we see the rainbow at all, when the overall intensity is so low? The answer to this question lies in the relationship between the angle ϕ and the angle of incidence, α. Let us again consider a large number of parallel rays, as in figure 6.30. However, there we have only drawn a section of the raindrop. To get the real picture, we have to imagine the raindrop in three dimensions. The light rays are then uniformly distributed over the whole cross-sectional area of the raindrop. Any two rays with different distances d_A and d_B to the symmetry axis of the raindrop will have different angles of incidence α and β. (See fig. 6.31.)

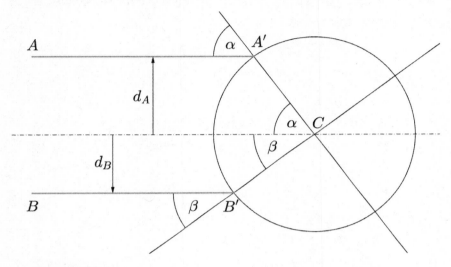

Figure 6.31.

The distances d_A and d_B are called the *impact parameters* of the rays AA' and BB'. If we let the configuration in figure 6.31 rotate around the axis of symmetry of the raindrop, we obtain two circles with radii d_A and d_B, representing the impact points of all rays that enter at an angle α or angle β, respectively. If d_A is twice as large as d_B, then the circumference of the circle with radius d_A will be twice the circumference of the circle with radius d_B. Hence, there will also be twice as many rays hitting the raindrop at an angle α than there will be rays entering at an angle β. This means that the distribution of light is not homogeneous over all possible angles of incidence. Most of the rays will have a comparatively large impact parameter, and, thus, will have a larger angle of incidence. We have already calculated the total deflection angle ϕ (see fig. 6.29) in terms of the angle of incidence α: A ray hitting the raindrop at an angle α will leave the raindrop in a direction such that the angle between the incident ray and the reflected ray is $4 \cdot \arcsin\left(\frac{\sin \alpha}{n}\right) - 2\alpha$.

This equation allows us to compute the total deflection angle $\phi = 4 \cdot \arcsin\left(\frac{\sin\alpha}{n}\right) - 2\alpha$ for any given angle of incidence. The graph in figure 6.32 shows the values of the total deflection angle for all possible angles of incidence (from 0° to 90°).

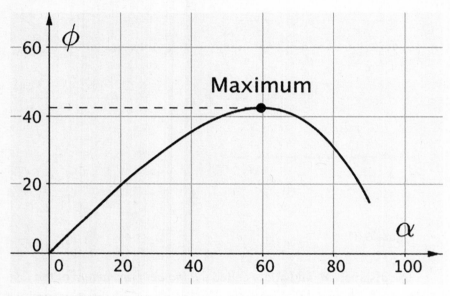

Figure 6.32.

We see that the curve representing ϕ as a function of α has a maximum at an angle of incidence $\alpha \approx 60°$ and a maximum deflection of $\phi \approx 42°$, meaning that ϕ cannot get larger than about 42°. Moreover, we can see from the diagram that rays with angles of incidence between 50° and 70° will leave the raindrop with an angle ϕ very close to 42°. Thus, the intensity of the outgoing rays will not be distributed homogeneously; it will be concentrated around an angle of 42°. A ray with the proper angle of incidence, such that the deflection angle ϕ exactly attains its maximally possible value, is called the "Descartes ray," since the French mathematician and philosopher René Descartes (1596–1650) was the first person who calculated this angle and provided a complete and correct explanation of the formation of rainbows.[15]

In figure 6.33 we show the whole interaction for a parallel bundle of incident rays coming in above the horizontal axis. All rays shown are refracted into the droplet, then reflected once inside the droplet, and finally refracted back into the air again. All rays emerging from the droplet are below the horizontal axis. We can see the concentration of intensity at the lower boundary of the bunch of outgoing rays, precisely at an angle of 42° with respect to the horizontal.

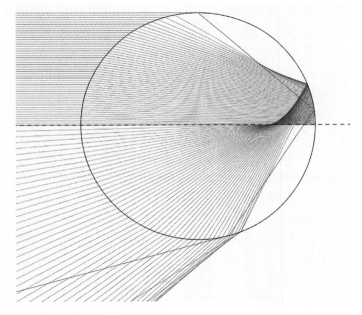

Figure 6.33.

However, in figure 6.33, we have omitted the rays coming in below the horizontal. To get the "full picture," the following two diagrams (figure 6.34) include the rays below the horizontal axis as well. This basically means to additionally reflect all the rays in figure 6.33 about the horizontal axis, leading to the picture in figure 6.34a. It shows both the incident rays and the rays returning after two refractions and one reflection. To better recognize the intensity peaks at 42° above

268 THE MATHEMATICS OF EVERYDAY LIFE

and below the horizontal, the picture in figure 6.34b shows only the returning rays, but not the incident rays.

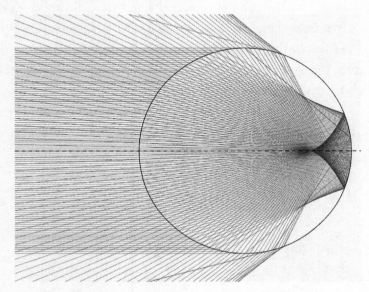

Figure 6.34a.

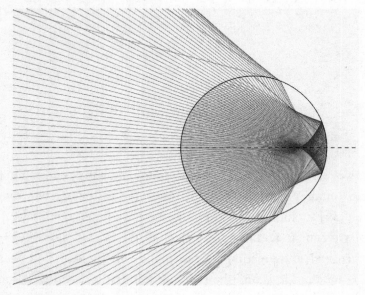

Figure 6.34b.

These diagrams show only a particular section through the droplet, defined by its center and the incident ray. For a three-dimensional picture, we have to imagine the formation that arises when we rotate the two-dimensional plots in figure 6.34 around the horizontal axis. Most of the rays emerging from the droplet will be concentrated on the boundary of a circular cone with half-angle 42°. The intensity in this direction is sufficiently high to make a rainbow visible. But where do we actually see the rainbow, and why is it colored? First of all, rain showers contain many droplets of water, at different heights and different points above the ground. Each droplet that catches sunlight will reemit light that is essentially confined to the boundary of a circular cone with half-angle 42°. To see this light sent out by the droplets, the light rays must strike our eyes. This means that our position in relation to the sun and the rain must be such that there exist view lines from our eyes to sunlit raindrops in the shower, which include an angle of 42° with a straight line drawn from the sun to the respective raindrops. Since all light rays from the sun are parallel to each other, this means that the same angle of 42° will also occur between a straight line drawn from the sun to the eye of the observer, and a view line from the observer's eye to a raindrop in the rainbow. This can be seen from figure 6.35. The collection of all view lines including an angle of 42° with a straight line drawn from the sun to the eye of the observer will form a circular cone with half-angle 42°, as can also be inferred from figure 6.35, where we have indicated part of this cone by dashed lines.

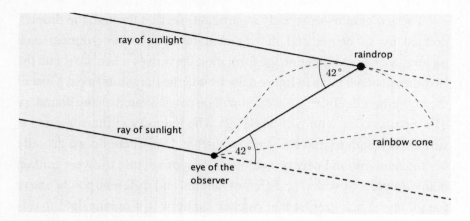

Figure 6.35.

It's exactly the light rays along this cone that produce the rainbow. A rainbow will always appear under an angle of 42°, measured from a straight line drawn from the sun through the eye of the observer. This line can also be identified as the line drawn from the observer's head to her or his head's shadow on the ground. Since we carry this 42° cone with us when we change our position on the ground, the rainbow we see will move with us. Its light will come from different positions in the rain shower. This explains why we can never reach a rainbow. The rainbow travels with us as we move. Different observers will see light coming from different positions within the rain shower and thus a different rainbow.

From figure 6.35, it is clear that rainbows actually form full circles, but when we are standing on the ground, we can only see light that emerges from raindrops above the horizon. We usually cannot see a rainbow's lower, hidden half. However, if you are flying in a plane toward a rain shower, with the sun at your back, you may be lucky enough to see a full-circle rainbow, since light from the raindrops can also reach you from below. If you are interested in seeing a full, circular rainbow, another option would be to climb on a tall mountain when the weather conditions are promising, and wait.

Why do rainbows have colors? Each ray of sunlight is actually composed of a whole spectrum of colors, as Isaac Newton (1643–1727) first noticed.[16] Newton also showed that red light has a slightly different refractive index in water compared to violet light, which is at the other end of the visible spectrum. The refractive index depends on the wavelength of light and is approximately $n = 1.332$ for red light and $n = 1.344$ for violet light.[17] For the other colors of the rainbow, the values lie in between these two extremes. Thus, the curve shown in the graph of figure 6.32 is slightly different for light of a different color. The maxima are attained at slightly different angles, implying that the high-intensity cones will not form at exactly 42° for all colors, but between approximately 40.5° (violet) and 42.2° (red light). Hence each color will form its own arc, and, thus, we see a whole band of colors. There are, in fact, millions of colors in a rainbow. The left picture in figure 6.36 shows the paths of the rays for four different colors.

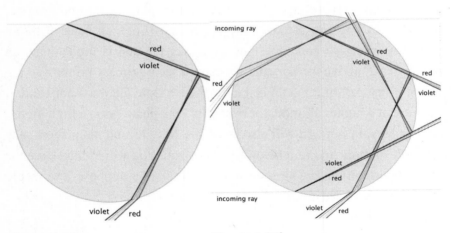

Figure 6.36a. Figure 6.36b.

The right picture shows how a secondary rainbow is formed. It is composed of light rays reflected twice inside the raindrop before they are refracted back into the air. Since each internal reflection reverses the order of the colors, the colors of a secondary rainbow are reversed as well. Secondary rainbows are much fainter than primary rainbows, since additional intensity is lost at the second reflection. They appear on the boundaries of cones with half-angles between about 52° and 54° with respect to a line drawn from the sun to the observer's eye, representing the apex of the cones.

Both the primary and the secondary rainbow are perfect circular arcs, centered at the antisolar point (see figure 6.37). The antisolar point is the imaginary point on the celestial sphere exactly opposite the sun from the viewpoint of an observer. Thus, if the sun is above the horizon, your antisolar point will be below the horizon. During sunrise or sunset, when the sun is just crossing the horizon, your antisolar point will also be crossing the horizon, exactly opposite to the sun. Since your antisolar point is always the center of the rainbow you are seeing, you will at most be able to see a half-circle (during sunrise or sunset). If you are standing on flat land, the hidden half of the rainbow would be below your horizon. However, there is neither sunlight nor rain below your horizon, so you can see only the upper half. Looking back at figure 6.35, we can also conclude that whenever the sun is higher than an angle of 42° above the horizon, we cannot see a primary rainbow at all, since every part of it would be below the horizon. On the other hand, if you are at a higher altitude, for example, on a plane or on a high mountain, then your horizon will also be at this altitude, and, thus, there may be both sunlight and rain below your horizon. You would then have the rare opportunity to see more than half a circle of a rainbow. You would even be able to see a full "raincircle."

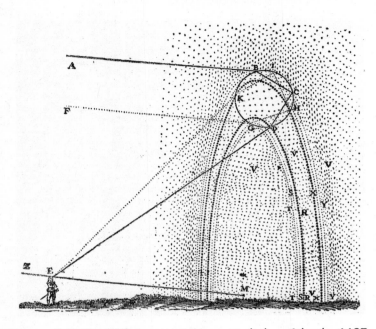

Figure 6.37. René Descartes, *Discours de la méthode*, 1637.

Of course, you have to be very lucky to experience a rainbow exceeding a half-circle in nature. But, as a matter of fact, it is not difficult to create an artificial rainbow in your garden, even one that is larger than a half-circle. You may have already done so as a child but have simply forgotten. You only need to climb a ladder, with the sun at your back, and turn on a garden hose. If you then adjust the hose's nozzle so the water comes out in a fine spray, a rainbow will appear in the spray. Moving up and down the ladder, the circular arc will increase and decrease, corresponding to your altitude.

We have now viewed and inspected how mathematics appears in nature. Our journey continues as we observe how man-made objects can be explained mathematically, specifically art and architecture.

CHAPTER 7

APPEARANCES OF MATHEMATICS IN ART AND ARCHITECTURE

When we apply mathematics to art and architecture, there is one overriding concept that seems to be omnipresent. It is the golden ratio, which is sometimes intentionally included, and, other times, it appears because various artists' perception of beauty in geometry tends to gravitate toward the rectangle shaped as the golden ratio. After we define the golden ratio, we will demonstrate its beauty and show its appearance in art and architecture. However, mathematics also can be seen in certain nonlinear examples, such as the catenary curve (a hanging chain supported on both ends at equal heights). We also will notice how the viewing perspective of art and in art is guided by mathematical principles. A drawing that exhibits the kind of perspective that allows us to appreciate depth perception is one such example, and there is also a mathematical justification for how we can stand to observe a sculpture or a painting in a museum. It needs to be understood that the mathematical appearances in art and architecture are practically limitless, and we make a small attempt to expose some of the more prominent examples here.

GOLDEN RATIO SIGHTINGS

It should not be so surprising that in art, as well as many other areas, mathematics has a way of manifesting itself. Since art has to be appealing to the human eye, the one overarching mathematical concept that seems to have positioned itself well in the arts and architecture is that of the golden ratio. Before exposing the appearances of this golden ratio we should define it. Geometrically, it is the ratio of the lengths of two line segments that allows us to make the following equality of two ratios:[1] that is, that the longer segment length (L) is to the shorter segment length (S) as the sum of the lengths of the segments ($L + S$) is to the longer segment length (L). Symbolically, this is written as $\frac{L}{S} = \frac{L+S}{L}$. (See fig. 7.1.)

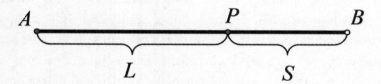

Figure 7.1.

To determine the numerical value of the golden ratio, $\frac{L}{S}$, we will change this equation[2] $\frac{L}{S} = \frac{L+S}{L}$, or $\frac{L}{S} = \frac{L}{L} + \frac{S}{L}$ to its equivalent, when $x = \frac{L}{S}$ to get the following:[3] $x = 1 + \frac{1}{x}$.

We can now solve this equation for x using the quadratic formula,[4] which you may recall from high school. We then obtain the numerical value of the golden ratio: $\frac{L}{S} = x = \frac{1+\sqrt{5}}{2}$, which is commonly denoted by the Greek letter, phi: ϕ.[5]

$$\phi = \frac{L}{S} = \frac{1+\sqrt{5}}{2} \approx \frac{1+2.2360679774997896964091736687312762354 40}{2}$$

$$\approx \frac{3.2360679774997896964091736687312762354 40}{2}$$

$$\approx 1.6180339887498948482045868343656381177 20$$

APPEARANCES OF MATHEMATICS IN ART AND ARCHITECTURE

Notice what happens when we take the reciprocal of $\frac{L}{S}$, namely, $\frac{S}{L} = \frac{1}{\phi}$:

$\frac{1}{\phi} = \frac{S}{L} = \frac{2}{1+\sqrt{5}}$, which, when we multiply by 1 in the form of $\frac{1-\sqrt{5}}{1-\sqrt{5}}$, we find that:

$$\frac{2}{1+\sqrt{5}} \cdot \frac{1-\sqrt{5}}{1-\sqrt{5}} = \frac{2 \cdot (1-\sqrt{5})}{1-5} = \frac{2 \cdot (1-\sqrt{5})}{-4} = \frac{1-\sqrt{5}}{-2} = \frac{\sqrt{5}-1}{2} \approx$$
0.6180339887498948482045868343656381177 20.

From the equation we got above, we can simplify it to obtain the quadratic equation as follows:

$x = 1 + \frac{1}{x}$ is equivalent to $x^2 - x - 1 = 0$.

(This equation also leads to the golden ratio: $x^2 + x - 1 = 0$.) To come full circle on this, we can see that the value ϕ satisfies the equation $x^2 - x - 1 = 0$, as shown here:

$$\phi^2 - \phi - 1 = \left(\frac{\sqrt{5}+1}{2}\right)^2 - \frac{\sqrt{5}+1}{2} - 1 = \frac{5+2\sqrt{5}+1}{4} - \frac{2(\sqrt{5}+1)}{4} - \frac{4}{4} =$$
$$\frac{5+2\sqrt{5}+1-2\sqrt{5}-2-4}{4} = 0.$$

The other solution of this equation is $\frac{1-\sqrt{5}}{2} = -\frac{\sqrt{5}-1}{2} = -\frac{1}{\phi}$, while $-\phi$ satisfies the equation $x^2 + x - 1 = 0$, as you can see here: $(-\phi)^2 + (-\phi) - 1 = \phi^2 - \phi - 1 = \left(\frac{\sqrt{5}+1}{2}\right)^2 - \frac{\sqrt{5}+1}{2} - 1 = 0$. The other solution to this equation is $\frac{1}{\phi}$.

At this point, you should notice a very unusual relationship. The values of ϕ and $\frac{1}{\phi}$ differ by 1. That is, $\phi - \frac{1}{\phi} = 1$. From the normal relationship of reciprocals, the product of ϕ and $\frac{1}{\phi}$ is also equal to 1, that is, $\phi \cdot \frac{1}{\phi} = 1$. Therefore, we have two numbers, ϕ and $\frac{1}{\phi}$, whose difference and product is 1—these are the *only* two numbers for which this is true! By the way, you might have noticed that $\phi + \frac{1}{\phi} = \sqrt{5}$, since $\frac{\sqrt{5}+1}{2} + \frac{\sqrt{5}-1}{2} = \sqrt{5}$.

Now that we have established the golden ratio, we can take it one step further to create a rectangle whose sides are in this golden ratio. To do that, we will begin with a unit square,[6] *ABCD*, with midpoint *M* of side *AB*, and then draw a circular arc *MC*, cutting the extension of side *AB* at point *E*. We now can claim that the line segment *AE* is partitioned into the golden ratio at point *B*. This, of course, has to be substantiated.

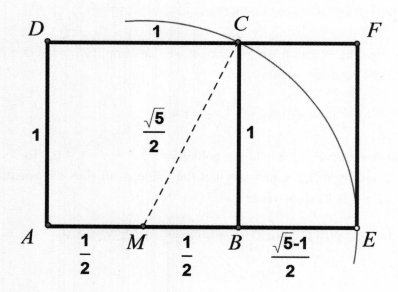

Figure 7.2.

To verify this claim, we would have to apply the definition of the golden ratio, $\frac{AB}{BE} = \frac{AE}{AB}$, and see if it, in fact, holds true. Substituting the values obtained by applying the Pythagorean theorem to $\triangle MBC$ as shown in figure 7.2, we get the following:

$$MC^2 = MB^2 + BC^2 = \left(\frac{1}{2}\right)^2 + 1^2 = \frac{1}{4} + 1 = \frac{5}{4}; \text{ therefore, } MC = \frac{\sqrt{5}}{2}.$$

APPEARANCES OF MATHEMATICS IN ART AND ARCHITECTURE 279

It follows that

$$BE = ME - MB = MC - MB = \frac{\sqrt{5}}{2} - \frac{1}{2} = \frac{\sqrt{5}-1}{2}, \text{ and}$$

$$AE = AB + BE = 1 + \frac{\sqrt{5}-1}{2} = \frac{2}{2} + \frac{\sqrt{5}-1}{2} = \frac{\sqrt{5}+1}{2}.$$

We then can find the value of $\frac{AB}{BE} = \frac{AE}{AB}$, that is, $\frac{1}{\frac{\sqrt{5}-1}{2}} = \frac{\frac{\sqrt{5}+1}{2}}{1}$, which turns out to be a true proportion, since the cross products are equal. That is, $\left(\frac{\sqrt{5}-1}{2}\right)\cdot\left(\frac{\sqrt{5}+1}{2}\right) = 1 \cdot 1 = 1$.

We can also see in figure 7.2 that point B can be said to divide the line segment AE into an inner golden ratio, since $\frac{AB}{AE} = \frac{1}{\frac{\sqrt{5}+1}{2}} = \frac{\sqrt{5}-1}{2} = \phi$.

Meanwhile, point E can be said to divide the line segment AB into an outer golden ratio, since $\frac{AE}{AB} = \frac{1+\frac{\sqrt{5}-1}{2}}{1} = \frac{\sqrt{5}+1}{2} = \phi$.

You ought to take notice of the shape of the rectangle $AEFD$ in figure 7.2. The ratio of the length to the width is the golden ratio: $\frac{AE}{EF} = \frac{\frac{\sqrt{5}+1}{2}}{1} = \frac{\sqrt{5}+1}{2} = \phi$. This appealing shape is called the *golden rectangle*.

For centuries, artists and architects have identified what they believed to be the most perfectly shaped rectangle. This ideal, *golden rectangle* has also proved to be the most pleasing to the eye. The desirability of this rectangle has been borne out by numerous psychological experiments. For example, Gustav Fechner (1801–1887), a German experimental psychologist, inspired by Adolf Zeising's book, *Der Goldene Schnitt*,[7] began a serious inquiry to see if the golden rectangle had a special psychological aesthetic appeal. His findings were published in 1876.[8] Fechner made thousands of measurements of commonly seen rectangles, such as playing cards, writing pads, books, windows, and so on. He found that most had a ratio of length to width that was close to ϕ. He also tested people's preferences and found most people preferred the shape of the golden rectangle.

280   THE MATHEMATICS OF EVERYDAY LIFE

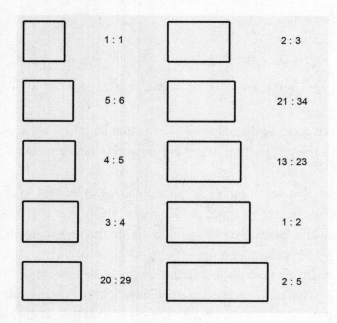

Figure 7.3.

What Gustav Fechner actually did was to ask 228 men and 119 women which of the following rectangles is aesthetically the most pleasing. Take a look at the rectangles in figure 7.3. Which rectangle would you choose as the most pleasing to look at? Rectangle 1:1 is too much like a square—considered by the general public as not representative of a "rectangle." It is, after all, a square! On the other hand, rectangle 2:5 (the other extreme) is uncomfortable to look at, since it requires the eye to scan it horizontally. The rectangle 21:34 can be appreciated at a single glance. Fechner's findings seem to bear this out. The results that Fechner reported are shown in table 7.1.

APPEARANCES OF MATHEMATICS IN ART AND ARCHITECTURE

Ratio of Sides of Rectangle	Percent Response for Best Rectangle	Percent Response for Worst Rectangle
1:1 = 1.00000	3.0	27.8
5:6 = .83333	.02	19.7
4:5 = .80000	2.0	9.4
3:4 = .75000	2.5	2.5
20:29 = .68966	7.7	1.2
2:3 = .66667	20.6	0.4
21:34 = .61765	35.0	0.0
13:23 = .56522	20.0	0.8
1:2 = .50000	7.5	2.5
2:5 = .40000	1.5	35.7
	100.00	100.00

Table 7.1.

Fechner's experiment has been repeated with variations in methodology many times, and his results have been further supported. For example, in 1917, the American psychologist and educator Edward Lee Thorndike (1874–1949) carried out similar experiments, with analogous results. In general, the rectangle with the ratio of 21:34 was most preferred. Do those numbers look familiar? Yes, once again, the Fibonacci numbers appear. The ratio $\frac{21}{34} = 0.6\overline{1764705882352941}$ approaches the value of ϕ and gives us the golden rectangle.

DISPLAYING A WATCH

We are not often aware of the many places that we experience aspects of the golden rectangle. For example, in almost all advertisements for watches and clocks, the hands of the clock show approximately 10:10.

Figure 7.4. Photo by R. N. Elliott.

The hands make an angle of $19\frac{1}{6}$ minutes, which is the equivalent of 115 degrees (fig. 7.4). (A quick aside to explain how we got this angle: Since 10 minutes is $\frac{1}{6}$th of an hour, the hour hand moved $\frac{1}{6}$ of the distance from "10" to "11" from 10:00 to 10:10. Therefore, the hour hand moved $\frac{1}{6}$ of 5 minute markers, or $\frac{5}{6}$ of a minute marker. Thus, the angle at 10:10 is $19\frac{1}{6}$ minute markers. To find the degree measure of that angle, we simply find the part of 60 minute markers that $19\frac{1}{6}$ is, in other words, $\frac{19\frac{1}{6}}{60} = \frac{\frac{115}{6}}{60} = \frac{115}{360} = 115°$.)

Now consider the rectangle formed by placing its vertices at points at which the hands indicate at 10:10, that is, at minute markers: 2 and $10\frac{1}{6}$. We can form a rectangle (fig. 7.5) with these two points as two adjacent vertices and have the point of intersection of its diagonals at the center of the watch face. This rectangle is very close to the golden rectangle, whose diagonals meet at an angle of about 116.6 degrees. (See fig. 7.6.)

APPEARANCES OF MATHEMATICS IN ART AND ARCHITECTURE 283

Figure 7.5

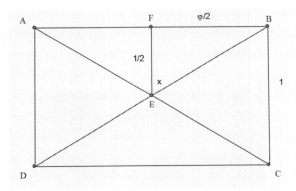

Figure 7.6.

This can be determined by using the golden rectangle (fig. 7.6) and finding the measure of the angle whose tangent function is $\tan x = \frac{FB}{FE} = \frac{\phi/2}{1/2} = \phi$. That angle, x, has measure 58.28 degrees. Therefore, $\angle AEB = 116.56°$, which is very close to the 115°, is felt by exhibitors to be the ideal placement of the hands of a watch. Some might have suspected that the 10:10 time placement was to allow the observer to see in unobstructed fashion the brand name of the watch. Not necessarily so, as is borne out of the above research, we simply prefer the appearance of the golden ratio and rectangle.

APPLICATIONS IN ART

The famous Italian painter, sculptor, architect, natural scientist, and engineer Leonardo da Vinci (1452–1519) illustrated the book *De divina proportione*[9] (*On the Divine Proportion*), which was written by the Franciscan monk Fra Luca Pacioli[10] (ca. 1445–1517). It included an anatomical study of the Vitruvian Man (fig. 7.7), Vitruvius Pollio (ca. 84 BCE–27 BCE), a Roman armed-forces technician and engineer who published ten books about architecture and civil engineering and developed a theory of proportions. Da Vinci included this illustration to support Pacioli's discussion of Vitruvius's documents about architecture, in which he represented the proportions of the human body as a the basis for architecture.

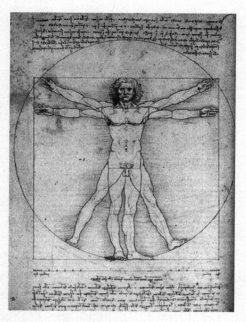

Figure 7.7. *Vitruvian Man*, by Leonardo da Vinci.

The ratio of square side to circle radius in this famous illustration corresponds to the golden ratio with a deviation of 1.7 percent.

In 1523, the famous German painter and graphic artist Albrecht Dürer (1471–1528) relied heavily on Vitruvius's work as he wrote his four books on human proportions. Here he refined the earlier works and expressed them as a system of proportions, where the unit of measure is the human body, with the parts expressed as fractions thereof.

In Dürer's books on geometry, fortress building, and human proportions, he supported all of his ideas with theoretical arguments and kept his artistic instincts aside. As a matter of fact, his works in descriptive geometry influenced some of the greatest thinkers of the Renaissance, including Johannes Kepler (1571–1630) and Galileo Galilei (1564–1642).

In his famous self-portrait (ca. 1500), Dürer drew himself with a head of wavy hair whose outlines create an equilateral triangle. This can be seen in figure 7.8, where Dürer actually superimposes the triangle and several other guidelines over the self-portrait. The base of the equilateral triangle divides a height of the entire picture into the golden ratio. The chin also divides the height of the entire picture into the golden ratio.

Figure 7.8. Albrecht Dürer.

Figure 7.9. Albrecht Dürer.

By the way, you will notice that when Dürer made his self-portrait (fig. 7.9), he made it so that his head could be encased perfectly into a golden rectangle. This is further evidence that Dürer was very aware of some of the techniques that da Vinci seemed to champion in his artwork: perspectivity and the golden ratio.

Sir Theodore Cook (1867–1928) analyzed Sandro Botticelli's (1445–1510) painting *The Birth of Venus* (1477)—shown in figure 7.10—with the aid of the golden ratio, whereby he placed a number scale on the figure of Venus, which had the first seven powers of ϕ. In many cases, we still do not know if the use of the golden ratio or the Fibonacci numbers was intentional or serendipitous. Many scholars have collected information in the nineteenth and twentieth centuries to search for applications of the golden ratio. These applications need to be critically and sensitively inspected. Oftentimes, we stumble on fractions that approach the golden ratio and immediately assume that the golden ratio was intended. We must be careful not to draw conclusions where they aren't necessarily warranted. Conscious use of the golden ratio often cannot be substantiated.

Figure 7.10. *The Birth of Venus*, Sandro Botticelli.

Whether architects and artists consciously use the golden ratio, when it appeared clearly in their work, was discussed by Dan Pedoe (1910–1998) in his book *Geometry and the Visual Arts*.[11] He argues that especially in Renaissance architecture, there is hardly any proof that the golden ratio was used intentionally, although it motivated great interest in the field. As they say, beauty is in the eye of the beholder, and so the beholder must make the judgment about the use of the golden ratio. We will concentrate our observation of paintings to a few samples where the golden ratio appears, yet there are countless additional examples that you can find in art and architecture over the past several centuries. Let us begin with perhaps one of the most famous paintings in Western civilization, Leonardo da Vinci's *Mona Lisa*, which was painted 1503–1506 and is on grand exhibit in the Louvre in Paris. (See fig. 7.11.)

Figure 7.11. *Mona Lisa*, Leonardo da Vinci.

Incidentally, King François I of France paid 15.3 kilograms of gold for this priceless masterpiece. To begin our inspection of the painting, we can draw a rectangle around Mona Lisa's face and find that this rectangle is a golden rectangle, as shown in figure 7.12.

In figure 7.13, you will notice several triangles; the two largest are golden triangles, which are isosceles triangles whose base and side are in the golden ratio.

Figure 7.12.

APPEARANCES OF MATHEMATICS IN ART AND ARCHITECTURE

Figure 7.13.

Furthermore, in figure 7.14, you will notice specific points on the body of Mona Lisa as golden ratios. Since da Vinci illustrated Pacioli's book *De divina proportione*, which thoroughly discussed the golden ratio, it can be assumed that he consciously was guided by this magnificent ratio.

Figure 7.14.

Raphael's (1483–1520) *Sistine Madonna* (1513) also exhibits the golden ratio. The horizontal line indicated in figure 7.15, which emanates from the eyes of Pope Sixtus II and the head of Saint Barbara, divides the height of the picture into the golden ratio and also partitions the figure of the Madonna into two equal parts. In figure 7.16, you will notice the white lines emanating from selected points in the figure in determining the golden ratio from another point of view, namely, the Madonna's figure is divided by the golden ratio. You will also notice the equilateral triangle superimposed on the picture in figure 7.16 exactly encases the four heads. Whether Raphael consciously selected these dimensions with the golden ratio in mind or whether these were merely the product of an artistic eye remains the master's secret.

Figure 7.15. *Sistine Madonna*, Raphael.

Figure 7.16.

In figure 7.17, the regular pentagram superimposed over Raphael's *Madonna Alba* (1511–1513), which is exhibited in the National Gallery, in Washington, DC, is clearly formed along appropriate linear parts and once again demonstrates Raphael's penchant for the golden ratio, which is embedded in the regular pentagram, as its sides are divided into the golden ratio.

Figure 7.17. *Madonna Alba*, Raphael.

Let's jump forward a few hundred years, to the French painter Georges Seurat (1859–1891), who was known for the strict geometric structure to his works. Although he used colors to a dramatic extent, Seurat wanted to highlight his use of geometric space as the main attraction and thereby set the groundwork for modern art.

292   THE MATHEMATICS OF EVERYDAY LIFE

Figure 7.18. *Circus Sideshow*, Georges Seurat.

Seurat's painting *Circus Sideshow* (1888), exhibited at the Metropolitan Museum of Art in New York, is replete with evidence of the golden ratio. Some observers see the golden ratio produced by the horizontal line at the top edge of the banister, and in the vertical direction by the line just to the right of the main figure. Others find that the rectangle formed by these two lines and the horizontal line just below the nine lights is a golden rectangle. The 8 × 3 unit rectangle (at the upper left side) gives further evidence to the possibility that the artist was aware of the relationship of the Fibonacci numbers to the golden ratio.

In Seurat's *Bathers at Asnières* (1883), exhibited at the Tate Gallery, in London, the golden ratio can be seen, as can several golden rectangles. (See fig. 7.19.)

APPEARANCES OF MATHEMATICS IN ART AND ARCHITECTURE 293

Figure 7.19. *Bathers at Asnières*, Georges Seurat.

Some artists, such as the Icelandic artist, Hreinn Friðfinnsson (1943–), clearly based their art on the golden rectangle and the golden spiral, which is traced by the quarter circles in the squares cut off from the golden rectangle. Friðfinnsson's piece called *Principle and Temptation* (1990), exhibited at the Museum of Art in Liechtenstein, Vaduz, can be seen in Figure 7.20, where each quarter circle cuts the prior radius in the golden ratio.

294   THE MATHEMATICS OF EVERYDAY LIFE

Figure 7.20. (Hreinn Friðfinnsson, *Principle and Temptation*, 1990–2013, C-print on aluminum, 133 × 200 cm; Courtesy of the artist and Galerie Nordenhake Berlin / Stockholm.)

As mentioned previously, the golden ratio can be found in many pieces of art and architecture. As such, we offer a list of works of art where there is some evidence that the golden ratio has been embedded in the composition of the work. Whether it was intentional or intuitive is open for discussion and research. Many of these can be found in books, online, or, for the adventurous reader, in person!

- The relief *Dionysius' Procession* (Villa Albani, Rome)
- The fresco *St. Francis Preaching to the Birds* by Giotto[12] (1266–1337) (Basilica San Francesco, Assisi, Italy)
- The fresco *Trinity* by Masaccio (1401–1429) (Santa Maria Novella, Florence, Italy)[13]
- The altarpiece *Deposition from the Cross* by Rogier van der Weyden[14] (ca. 1400–1464) (Prado, Madrid)
- The central panel of a polyptych *The Baptism of Christ* by Piero della Francesca[15] (1415/1420–1492) (National Gallery, London)

APPEARANCES OF MATHEMATICS IN ART AND ARCHITECTURE 295

- The oil painting *Madonna and Child* by Pietro Perugino[16] (1445/1448–1523)
- The painting *The Girl with the Ermine* by Leonardo da Vinci (National Museum, Krakow, Poland)
- The wall painting *The Last Supper* by Leonardo da Vinci (Refectory of Santa Maria delle Grazie, Milan)
- The round painting of *The Doni Tondo*, also called *The Holy Family with St. John the Baptist*, by Michelangelo[17] (1475–1564) (Uffizi Gallery, Florence)
- The painting *Crucifixion* (with the Virgin Mary, Saint Jerome, Mary Magdalene, and John the Baptist) by Raphael (1483–1520) (National Gallery, London)
- The panels *Adam and Eve* by Albrecht Dürer (Prado Museum, Madrid)
- The painting *School of Athens* by Raphael (Vatican museum, Rome)
- The fresco *The Triumph of Galatea* by Raphael (Villa Farnesina, Rome)
- The copper engraving *Adam and Eve* by Marcantonio Raimondi[18] (1480–ca. 1530–34) (Harvard Art Museums/Fogg Museum)
- The painting *A Self-Portrait* by Rembrandt Harmenszoon van Rijn (1606–1669) (National Gallery, London)
- The oil painting on linen *Gelmeroda* by Lyonel Feininger (1871–1956) (*Gelmeroda* VIII, Whitney Museum, New York; and *Gelmeroda* XII, Metropolitan Museum of Art, New York)

Further, the following is a partial list of artists who are often mentioned in connection with the golden ratio:

- Paul Signac (1863–1935)
- Paul Sérusier (1864–1927)
- Piet Mondrian[19] (1872–1944)

- Juan Gris[20] (1887–1927)
- Otto Pankok[21] (1893–1966)

While it is known that Paul Sérusier not only knew about the golden ratio but also indicated it in his sketches, Gris, Mondrian, and Pankok have flatly denied using it.

Many admirers of the golden ratio and the Fibonacci numbers would be disappointed to learn that the careful examination by the art historian Marguerite Neveux at the end of the twentieth century removed many pictures from the "golden ratio list." She analyzed x-ray pictures of various canvases and came to the conclusion that most of the artists divided their canvas into eighths before starting their work. There are many ways to use these fractional partitions; yet, more often than not, $\frac{5}{8}$ is selected. So, if this art historian eliminated some art from the "golden ratio list," we can still claim them to the "almost golden ratio list" as they used two consecutive Fibonacci numbers that can generate a rough approximation of the golden ratio. It is truly fascinating how this "magical" golden ratio has attracted artists over the centuries—sometimes deliberately and other times intuitively. Oftentimes, it is for the viewer to make the judgment of its appearance.

PERSPECTIVITY IN ART

The challenge of capturing perfect depth perception in drawings is often attributed to the earliest artists of the Renaissance. Most credit for initiating the study is attributed to the Italian architect and designer Filippo Brunelleschi (1377–1446) and to the Italian writer and artist Leon Battista Alberti (1404–1472), who with their writings and artistic renditions opened new vistas for themselves and future artists as well. This can be best seen through some of the most famous early examples of this technique.

APPEARANCES OF MATHEMATICS IN ART AND ARCHITECTURE 297

Over the centuries there have been many renditions of *The Last Supper* by famous painters. As time went on, these artists pursued attempts at capturing proper depth perception in their paintings. A technique in art that allows perfect depth perception is referred to as *perspectivity*. It is believed that of the many renditions of *The Last Supper* paintings, the first one that captured true and complete perspective, was the famous Italian artist Leonardo da Vinci, with his mural in the Santa Maria delle Grazie in Milan. We see the lines of perspectivity in figure 7.21, surrounding Jesus's head. Actually, when the lines are extended to where they all meet, studies have shown that they converge on Jesus's right eye, which could then be considered the *center of perspectivity*.

Figure 7.21. *The Last Supper*, Leonardo da Vinci.

In figure 7.22, we show the various lines that allowed Leonardo da Vinci to capture true perspectivity in his painting.

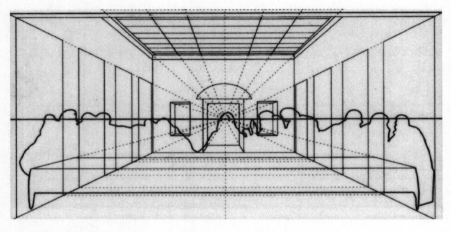

Figure 7.22.

Figure 7.23. *The Virgin of the Rocks*, Leonardo da Vinci.

Leonardo da Vinci kept a record of his work in many notebooks. There he described three factors that also affected perspective appearance. These were: color, size, and disappearance, which could be considered objects somewhat less sharp in the rear of the picture than those in front. One illustration of this is his 1508 painting *The Virgin of the Rocks*, which is shown in figure 7.23.

Before exploring other masterpieces, let's digress and understand what perspectivity means from a purely geometrical standpoint. In figure 7.24, we say that the segment AB and the segment DC are in perspective at point P, which is their center of perspectivity. The term *perspective* was likely introduced from optics, since the eye placed at P would see the point D coinciding with the point A, and the point C coinciding with the point B. Furthermore, the line segment DC would appear to coincide with the line segment AB. This will allow us to understand this concept as it has been used by artists to create their masterpieces.

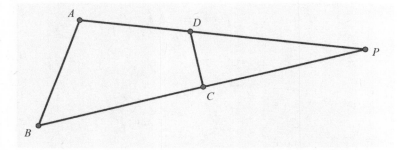

Figure 7.24.

It is well-known that Leonardo da Vinci spent a lot of time analyzing his artwork, which can be seen very clearly with his perspectivity study of *The Adoration of the Magi*. This is shown in figure 7.25, where we notice the lines of perspective are drawn in by the artist himself.

Figure 7.25. *The Adoration of the Magi*, Leonardo da Vinci.

Around the year 1510, the Italian artist Raphael painted his famous *School of Athens* (fig. 7.26), where he also employed the concept of perspectivity.

Figure 7.26. *School of Athens*, Raphael.

We show the lines of perspectivity in figure 7.27. Once again, we notice how with this technique perfect depth perception is achieved in the artwork.

Figure 7.27. *School of Athens*, Raphael.

APPEARANCES OF MATHEMATICS IN ART AND ARCHITECTURE

About the same time Leonardo da Vinci created his masterpieces, the famous German artist Albrecht Dürer also experimented with perspective in his drawings. In 1506, he made a special trip to Bologna, Italy, with the intent to learn about the secret theoretical foundation of a process that he already had seemed to master, namely, capturing perspectivity in his artwork. In figure 7.28 we notice how Dürer very consciously was aware of the concept of perspective. This was further substantiated when he published his manual in 1525, *Underweysung der Messung* ("Instruction in Measurement"), which was a treatise on the concept of perspectivity in geometry. This was targeted at well-educated artists and was followed by other such manuals for less well-prepared artists by other experts, such as Hieronymus Rodler (?–1539), who in 1531 published a simpler version titled *Eyn schön nützlich Büchlein und Underweisung der Kunst des Messens* ("A Nice Useful Booklet and Instruction of the Art of Measurement").

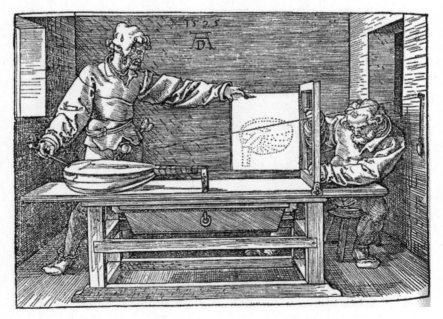

Figure 7.28. *Man Drawing a Lute*, Albrecht Dürer.

We can see the actual effect of his consciousness with using perspective to draw his famous etching *St. Jerome in His Study*, shown in figure 7.29.

Figure 7.29. *St. Jerome in His Study*, Albrecht Dürer.

In a painting, a more realistic portrayal of a scene can also be achieved by creating the illusion of light. Playing with light and shadow

APPEARANCES OF MATHEMATICS IN ART AND ARCHITECTURE 303

can significantly enhance perspective appearance. A wonderful example of this is the masterpiece *The Calling of Saint Matthew* by Michelangelo Merisi da Caravaggio (1571–1610), which is shown in figure 7.30. For a convincing illusion of light falling through the window, it is important that the artist is aware of the correct shape of cast shadows. Drawing cast shadows realistically involves using the geometric principles of perspective drawing, since the boundary of a shadow can be constructed by using perspective lines emanating from the light source.

Figure 7.30. *The Calling of Saint Matthew*, Caravaggio.

When we regard the depiction of a real object on a painting, we basically see a projection of the object onto a plane that is represented by the canvas. If this projection is distorted, the viewer might have to occupy a very specific vantage point to reconstitute the image. This is called *perspectival anamorphosis*, and it requires the artist to master the laws of perspective geometry. While Leonardo da Vinci was probably the first to experiment with distorted images that appear in natural form when viewed at a certain angle, perhaps the most famous example of perspectival anamorphosis can be found in the painting *The Ambassadors*, by Hans Holbein the Younger (1497–1543), which is exhibited in the National Gallery in London. A highly distorted shape at the bottom will turn into the plastic image of a skull when viewed from an acute angle. Although by far not as impressive as the original, you may convince yourself of this remarkable effect by looking at the photograph of Holbein's painting shown in figure 7.31 from a different angle until the seemingly amorphous shape toward the bottom of the image eventually becomes a perfectly rendered painting of a skull (angle the book away from you, keeping the left side closer to your body and the right side farther from your body).

Figure 7.31. *The Ambassadors*, Hans Holbein the Younger.

There are countless other examples of perspectivity used in art, especially where the artist wanted to accurately capture depth perception. Once again, mathematics seems to play a significant role where we might not have expected it to.

NUMBERS IN ART

We can also revisit the renowned German artist Albrecht Dürer (1471–1528), who lived in Nuremberg, Germany, and, aside from his art, has also provided us with a rather-unusual number arrangement called a *magic square*. Magic squares are square arrangements of numbers for which the sum of each of the columns, rows, and diagonals are all equal. However, there is one magic square that stands out among the rest for its beauty and *additional* properties—not to mention its curious appearance. This particular magic square has many properties beyond those required for a square arrangement of numbers to be considered "magic." This magic square even comes to us through art, and not through the usual mathematical channels. It is depicted in the background of Dürer's famous 1514 engraving *Melencolia I* (see fig. 7.32).

Figure 7.32. *Melencolia I*, Albrecht Dürer, 1514.

306 THE MATHEMATICS OF EVERYDAY LIFE

As we begin to examine the magic square in Dürer's etching, we should take note that most of Dürer's works were signed by him with his initials, one over the other, and with the year in which the work was made included. Here we find it in the dark-shaded region near the lower right side of the picture (see figs. 7.32 and 7.33). We notice that it was made in the year 1514.

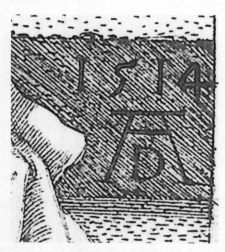

Figure 7.33. Initials of Albrecht Dürer, AD, and the year, 1514.

Consider now the magic square located in *Melencolia I*. The observant reader may notice that the two center cells of the bottom row of the Dürer magic square depict the year as well. Let us examine this magic square more closely. (See fig. 7.34.)

APPEARANCES OF MATHEMATICS IN ART AND ARCHITECTURE

 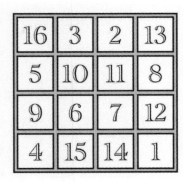

Figure 7.34. Dürer's magic square.

First, let's make sure that it is, in fact, a true magic square. When we evaluate the sum of each of the rows, columns, and diagonals, we get the result 34. This feature alone is all that would be required for this square matrix of numbers to be considered a "magic square." However, this Dürer magic square has many more properties that other magic squares do not have. Let us now marvel at some of these extra properties.

- The four corner numbers have a sum of 34:
    16 + 13 + 1 + 4 = 34
- Each of the four corner 2 × 2 squares has a sum of 34:
    16 + 3 + 5 + 10 = 34
    2 + 13 + 11 + 8 = 34
    9 + 6 + 4 + 15 = 34
    7 + 12 + 14 + 1 = 34
- The center 2 × 2 square has a sum of 34:
    10 + 11 + 6 + 7 = 34
- The sum of the numbers in the diagonal cells equals the sum of the numbers in the cells not in the diagonal:
    16 + 10 + 7 + 1 + 4 + 6 + 11 + 13 = 3 + 2 + 8 + 12 + 14 + 15 + 9 + 5 = 68

- The sum of the squares of the numbers in both diagonal cells is as follows:
  $$16^2 + 10^2 + 7^2 + 1^2 + 4^2 + 6^2 + 11^2 + 13^2 = 748$$
  This number is equal to
  - the sum of the squares of the numbers not in the diagonal cells:
    $$3^2 + 2^2 + 8^2 + 12^2 + 14^2 + 15^2 + 9^2 + 5^2 = 748$$
  - the sum of the squares of the numbers in the first and third rows:
    $$16^2 + 3^2 + 2^2 + 13^2 + 9^2 + 6^2 + 7^2 + 12^2 = 748$$
  - the sum of the squares of the numbers in the second and fourth rows.
    $$5^2 + 10^2 + 11^2 + 8^2 + 4^2 + 15^2 + 14^2 + 1^2 = 748$$
  - the sum of the squares of the numbers in the first and third columns.
    $$16^2 + 5^2 + 9^2 + 4^2 + 2^2 + 11^2 + 7^2 + 14^2 = 748$$
  - the sum of the squares of the numbers in the second and fourth columns.
    $$3^2 + 10^2 + 6^2 + 15^2 + 13^2 + 8^2 + 12^2 + 1^2 = 748$$
- The sum of the cubes of the numbers in the diagonal cells equals the sum of the cubes of the numbers not in the diagonal cells:
  $$16^3 + 10^3 + 7^3 + 1^3 + 4^3 + 6^3 + 11^3 + 13^3 =$$
  $$3^3 + 2^3 + 8^3 + 12^3 + 14^3 + 15^3 + 9^3 + 5^3 = 9{,}248$$
- Further, notice the following beautiful symmetries:
  $$2 + 8 + 9 + 15 = 3 + 5 + 12 + 14 = 34$$
  $$2^2 + 8^2 + 9^2 + 15^2 = 3^2 + 5^2 + 12^2 + 14^2 = 374$$
  $$2^3 + 8^3 + 9^3 + 15^3 = 3^3 + 5^3 + 12^3 + 14^3 = 4624$$
- Adding the first row to the second and the third row to the fourth produces a pleasing symmetry:

16 + 5 = 21	3 + 10 = 13	2 + 11 = 13	13 + 8 = 21
9 + 4 = 13	6 + 15 = 21	7 + 14 = 21	12 + 1 = 13

- Adding the first column to the second and the third column to the fourth produces a pleasing symmetry:

16 + 3 = **19**	2 + 13 = **15**
5 + 10 = **15**	11 + 8 = **19**
9 + 6 = **15**	7 + 12 = **19**
4 + 15 = **19**	14 + 1 = **15**

A motivated reader may wish to search for other patterns in this beautiful magic square. Remember, this is not a typical magic square for which all that is required is that all the rows, columns, and diagonals have the same sum. This Dürer magic square has many more properties!

VIEWING A STATUE OPTIMALLY

Most cities in the world exhibit some magnificent statues. The statues are often on very high pedestals, which forces us to look up at them, especially when we are very close to the statue. Of course, as we move away from the statue, we get a better picture of the person depicted therein. However, there must be an optimal position at which we can stand to view the full statue in the best way possible.

To analyze this situation, let us begin with a simple geometric rendition of how this would play out. In figure 7.35, we will place our eye level at point A, and the actual statue will be depicted by the segment BD (s). The segment DC (p) will represent the height of the pedestal above our eye level.

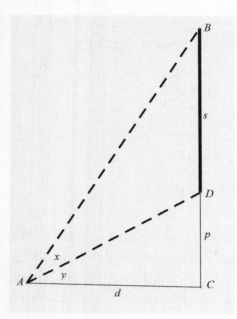

Figure 7.35.

To begin our quest for the best place to stand, we will first need to find the tangent functions as follows:

$\tan \angle BAC = \tan(x+y) = \frac{s+p}{d}$
$\tan \angle DAC = \tan y = \frac{p}{d}$

We then use a formula from high-school trigonometry for the tangent function of an angle sum:

$\tan \angle BAC = \tan(x+y) = \frac{\tan x + \tan y}{1 - \tan x \cdot \tan y}$, which yields upon solving for $\tan x$:

$\tan x = \frac{\tan(x+y) - \tan y}{1 + \tan(x+y) \cdot \tan y}$

Applying this formula to figure 7.35, we substitute $\tan(x+y) = \frac{s+p}{d}$ and $\tan y = \frac{p}{d}$ to find the value $\tan x$:

$$\tan x = \frac{\tan(x+y) - \tan y}{1 + \tan(x+y) \cdot \tan y} = \frac{\frac{s+p}{d} - \frac{p}{d}}{1 + \frac{s+p}{d} \cdot \frac{p}{d}} = \frac{\frac{s}{d}}{1 + \frac{p(s+p)}{d^2}} = \frac{sd}{p(s+p) + d^2}$$

We want to find the value for d, the distance from the base of the pedestal to the eye of the observer, which will give us the largest angle BAD for given p and s. This will optimize our view of the statue. Of course, the largest angle occurs when its tangent function, $\tan x$, is largest. Thus, we have to maximize $\frac{sd}{p(s+p)+d^2}$ for fixed p and s. This is equivalent to minimizing the reciprocal $\frac{p(s+p)+d^2}{sd} = \frac{1}{s}\left(\frac{p(s+p)}{d} + d\right)$, which is equivalent to finding the value for d, which will give the smallest value for $\frac{p(s+p)}{d} + d$. Now we use a little trick and write this as

$$\left(\sqrt{\frac{p(s+p)}{d}} - \sqrt{d}\right)^2 + 2\sqrt{p(s+p)},$$

a technique known as "completing the square." Obviously, this sum attains its smallest possible value when the value of $\left(\sqrt{\frac{p(s+p)}{d}} - \sqrt{d}\right)$ is equal to 0, which is the case if $\sqrt{\frac{p(s+p)}{d}} = \sqrt{d}$, which simplifies to $p(s+p) = d^2$. Or, to put it another way, the largest angle BAD occurs when $d = \sqrt{p(s+p)}$, which represents the geometric mean between the height of the pedestal (p) and the top height of the statue and the pedestal ($s + p$).

Suppose we try this with perhaps one of the most famous statues in Western culture, the statue of *David*, by Michelangelo. The statue is 5.17 meters high, and we will assume that the pedestal above eye level is 1 meter high. Therefore, using our formula, $d = \sqrt{p(s+p)} = \sqrt{1(5.17)+1} = 2.48$, which determines the distance we need to stand back from the statue to see it optimally (2.48 meters).

Another example of this technique for finding the best place to view something is when a museum puts up an exhibit and hangs big pictures on the wall; the question often arises, How high should the picture be hung so that a person sitting on a bench at a specific distance from the wall would be able to have the largest viewing angle of this picture?

312 THE MATHEMATICS OF EVERYDAY LIFE

This calculation is done exactly as we have done with the statue above. Simple geometry can be very useful for practical issues.

THE MOST OVERLOOKED CURVE

One of the most pervasive curves in our environment is the curve that holds up a suspension bridge. Two classic examples are on either side of the United States. On the West Coast, we have the famous Golden Gate Bridge, as shown in figure 7.36. And on the East Coast, we have the famous George Washington Bridge, shown in figure 7.37.

Figure 7.36.

Figure 7.37.

Although these curves very closely resemble one of the most popular curves we learn about in our high-school mathematics course, the parabola, the curve that holds up suspension bridges is *not* a parabola. In 1683 the famous Italian scientist Galileo Galilei (1564–1642) claimed that this curve, also seen as a hanging chain supported by two vertical poles, is, in fact, a parabola. The German mathematician Joachim Jungius (1587–1657) in 1639 showed Galileo to be wrong.[22] The actual equation describing this curve was finally established in 1691 largely by the Dutch mathematician Christiaan Huygens (1629–1695), who in a letter to a collaborator, the German mathematician Gottfried Leibniz (1646–1716), referred to this curve as a "catenaria," which was derived from the Latin word *Catena*, which means "chain."[23] The current Anglicized word *catenary* seems to be attributed to United States President Thomas Jefferson in a letter to Thomas Paine dated September 15, 1788, in which he described the design of the bridge.[24] In addition to Leibniz and Huygens, Swiss mathematician Johann Bernoulli (1667–1748) helped develop the formula for the catenary curve. When a chain is suspended from two equally high supports and is dropping at $x = 0$ to its lowest height $y = a$, its height y as a function of x is given by the equation: $y = \left(\frac{a}{2}\right)\left(e^{\frac{x}{a}} + e^{-\frac{x}{a}}\right)$, or, at a more advanced level (and also in the *xy*-plane), it can be expressed in terms of a hyperbolic cosine function as $y = a \cosh\left(\frac{x}{a}\right)$. This formula will still hold for different distances between the two poles holding a hanging chain, however, for different values of a.

The same formula will also hold if the supports are not equally high. In this case, the lowest height of the chain will occur closer to the shorter of the two supports. The formula applies for any chain supported only at its ends, as long as the *x*-coordinate in the equation is measured from the point of lowest height, a. We can also see a catenary in its most simple form, a hanging chain as we show in figure 7.38.

Figure 7.38. (Image from Wikimedia Creative Commons, author: Kamel 15; licensed under CC BY-SA 3.0 Unported.)

If we invert the catenary, we can create a stable, self-supporting arch, which is evidenced by the famous Gateway Arch in St. Louis, Missouri, which is shown in figure 7.39.

Figure 7.39. (Image from Wikimedia Creative Commons, author: Bev Sykes from Davis, CA; licensed under CC BY-SA 2.0.)

APPEARANCES OF MATHEMATICS IN ART AND ARCHITECTURE

So now you can see that a very often-observed curve that most people (including Galileo) thought was a parabola, in fact, is not, and yet we do understand its properties, thanks to the work of many brilliant mathematicians.

THE ONE-SIDED BELT—THE MÖBIUS STRIP

Surely you have seen the symbol shown in figure 7.40. This ubiquitous recycling symbol has been around for decades and decorates many products of a variety of sorts. The symbol is a somewhat-artistic version of the famous Möbius strip, and it was created in 1970 by Gary Anderson, a student at the University of Southern California, to symbolize recycling.[25]

Figure 7.40.

Merely as an artistic descriptor, this structure would have no real significance for us in mathematics. However, the Möbius strip has a very significant role in industry and beyond. For example, when we observe a fan belt being used in an automobile engine, as well as in some industrial machinery, the belt after time will wear out on the inside surface. Imagine a belt that does not have an inside or an outside surface but merely has one single surface. Such a belt was discovered independently by the two German mathematicians August Ferdinand Möbius (1790–1868) and Johan Benedict Listing (1808–1882) in 1858. You might ask yourself,

How can a belt have only one side? You can easily construct one by taking a strip of paper and giving it one twist before connecting the two ends, and you will get a loop that looks like that shown in figure 7.41.

Figure 7.41. (Image from Wikimedia Creative Commons, author: David Benbennick, own work; licensed under CC BY-SA 3.0.)

To convince yourself that this strip has only one side, take a pencil and run the pencil along the center of the surface without lifting it up. Eventually, you will find that the line you are drawing will connect to the point at which you began. Therefore, a fan belt in the form of a Möbius strip will have only one surface, which would be twice as long as the fan belt would have been without that twist. It will last longer than a conventional fan belt, where only one side of the belt would be in contact with the pulleys and thus get worn out sooner.

There are some curious properties of such a strip of paper. Suppose with a pair of scissors you cut the strip along the line you drew halfway from the edge. The result will be a long strip with two twists in it, which is *not* a Möbius strip. The reason for this is that the original strip had only one edge, and now you created a new strip with two edges, which was twice as long as the original strip. If we continue this cutting process, by cutting along the center of this new strip, we will end up with two strips interlocking, with each having two full twists.

Suppose we take our original Möbius strip and this time we cut along a line that is one-third of the distance from the edge. This time we will form two new strips: one will be a Möbius strip, which will have one-third the width of the original Möbius strip, yet it will have the same length as the original Möbius strip. The other strip will be twice as long as the original Möbius strip and will have two twists in it.

When an odd number of twists to the original strip of paper is made, a form of Möbius strip results, but this does not happen for an even number of twists. An interested reader may wish to experiment by creating various versions of the strips and cutting them at different distances from the edge. The results can be quite astonishing!

The famous Mexican sculptor Sebastián (Enrique Carbajal, 1947–) created a three-dimensional Möbius strip, which we show in figure 7.42.

Figure 7.42.

Here we see a geometric marvel that shows itself as a consumer symbol, an efficient component in industry, as well as a manifestation in the arts. As you can see from our discussion, it also has some entertaining aspects and surprises that bring some enjoyment from the field of mathematics.

Having journeyed through an artistic world dominated by mathematical occurrences and explanations, we now embark on a discussion of where we find mathematics helping us understand our structural environment from a mathematical and technological point of view.

CHAPTER 8

THE TECHNOLOGY AROUND US—FROM A MATHEMATICAL PERSPECTIVE

In today's world we can hardly escape being technologically involved. However, too often we just take for granted the technology around us. We observe the workings of a clock without thinking about its "timing." We appreciate timed traffic lights without thinking about how this works. We are amazed at the effects of a whispering spot in a large hall, and then we wonder why that works. These are just a few of the technology aspects that we encounter in our daily experiences that can be explained through simple mathematics. Let's continue now by considering these and a few other such encounters with "hidden" mathematics.

A FASCINATION WITH THE CLOCK

It is not uncommon to look at the clock many times during the day. When we look at a clock, we might wonder at what time the hands of a clock overlap, at what time they form an exact right angle, or, for that matter, at what time they form a straight angle. We do that by mere observation, but mathematics provides us with a very simple technique for determining the answers to these questions. For instance, at what time (exactly) will the hands of a clock overlap after four o'clock?

A typical first reaction to answer this question would be that the answer is simply 4:20. When you reconsider that the hour hand moves uniformly, you may begin to estimate the answer to be between 4:21 and 4:22. The hour hand moves through an interval between minute markers every 12 minutes. Therefore, it will leave the interval 4:21–4:22 at 4:24. This, however, doesn't answer the original question about the exact time of this overlap.

The best way to begin to understand the movement of the hands of a clock is by considering the hands traveling independently around the clock at uniform speeds. The minute markings on the clock (from now on referred to as "markers") will serve to denote distance as well as time. An analogy should be drawn here to the uniform motion of automobiles (a popular and overused topic for verbal problems in an elementary algebra course). That is, the hands of the clock can be seen analogously to a fast automobile (the minute hand) overtaking a slower automobile (the hour hand).

It might be helpful to find the distance necessary for a car traveling at 60 mph to overtake a car with a head start of 20 miles and traveling at 5 mph. (The m here represents "miles" but could just as easily represent "markers.") Now consider four o'clock as the initial time on the clock. Our problem will be to determine exactly when the minute hand will overtake the hour hand after four o'clock. Consider the speed of the hour hand to be r, then the speed of the minute hand must be $12r$. We seek the distance, measured by the number of markers traveled, that the minute hand must travel to overtake the hour hand. Let us refer to this distance as d markers. Hence, the distance that the hour hand travels is $d - 20$ markers, since it has a 20-marker head start over the minute hand. For this to take place, the time required for the minute hand, $\frac{d}{12r}$, and for the hour hand, $\frac{d-20}{r}$, would have to be the same. Therefore, $\frac{d}{12r} = \frac{d-20}{r}$, and then $d = \frac{12}{11} \cdot 20 = 21\frac{9}{11}$. Thus, the minute hand will overtake the hour hand at exactly $4:21\frac{9}{11}$.

Consider the expression $d = \frac{12}{11} \cdot 20$. The quantity 20 is the number

THE TECHNOLOGY AROUND US

of markers that the minute hand had to travel to get to the desired position, assuming the hour hand remained stationary. However, quite obviously, the hour hand does not remain stationary. Hence, we had to multiply this quantity by $\frac{12}{11}$, since the minute hand must travel $\frac{12}{11}$ as far. Let us refer to the fraction $\frac{12}{11}$ as the *correction factor*.

To become familiar with the correction factor, let's consider a few simple problems. For example, you choose to find the exact time when the hands of a clock overlap between seven and eight o'clock. Here we would first determine how far the minute hand would have to travel from the "12" position to the position of the hour hand, assuming again that the hour hand remains stationary. Then, by multiplying the number of markers, 35, by the correction factor, $\frac{12}{11}$, we will obtain the exact time that the hands will overlap: $7:38\frac{2}{11}$.

To enhance our understanding of this procedure, consider a person checking a wristwatch against an electric clock and noticing that the hands on the wristwatch overlap every 65 minutes (as measured by the electric clock). We then ask ourselves whether the wristwatch is fast, slow, or accurate. To do this, you may wish to consider the problem in the following way. At twelve o'clock, the hands of a clock overlap exactly. Using the previously described method, we find that the hands will again overlap at exactly $1:05\frac{5}{11}$, then at exactly $2:10\frac{10}{11}$, then at exactly $3:16\frac{4}{11}$, and so on. Each time, there is an interval of $65\frac{5}{11}$ minutes between overlapping positions. Hence, the person's watch is inaccurate by $\frac{5}{11}$ of a minute. Then is the wristwatch fast or slow?

There are many other interesting, and sometimes rather-difficult, problems that can be made simple by this correction factor. You may very easily pose your own problems. For example, you may wish to find the exact times when the hands of a clock will be perpendicular (or form a straight angle) between, say, eight and nine o'clock. Again, you would need to determine the number of markers that the minute hand would have to travel from the "12" position until it forms the desired angle with the stationary hour hand. Then, simply multiply this number by the correction factor of $\frac{12}{11}$ to obtain the exact actual time. That is, to find the exact time that the hands of a clock are *first* perpendicular between eight and nine o'clock, determine the desired position of the minute hand when the hour hand remains stationary (here, on the 25-minute marker). Then, multiply 25 by $\frac{12}{11}$ to get $8:27\frac{3}{11}$, which is the exact time when the hands are *first* perpendicular after eight o'clock.

One might also justify the correction factor $\frac{12}{11}$ for the interval between overlaps in the following way: Think of the hands of a clock at noon. During the next 12 hours (that is, until the hands reach the same position at midnight) the hour hand makes one revolution, the minute hand makes 12 revolutions, and the minute hand coincides with the hour hand 11 times (including midnight, but not noon, starting just after the hands separate at noon). Because each hand rotates at a uniform rate, the hands overlap each $\frac{12}{11}$ of an hour, or $65\frac{5}{11}$ minutes. This can be extended to other situations, which can prove to be fun while at the same time showing how mathematics can be applied to the travels of the hands of a clock.

THE MATHEMATICS OF PAPER FOLDING

Take a piece of paper, fold it in half, then fold it in half again, and keep folding the paper in half. How many times can you fold it before it won't fold any further? It is a counterintuitive fact that it is impossible to fold a piece of paper in half more than seven or eight times. With each additional folding, the thickness of the folded paper doubles, while its surface area is halved (see fig. 8.1).

THE TECHNOLOGY AROUND US 323

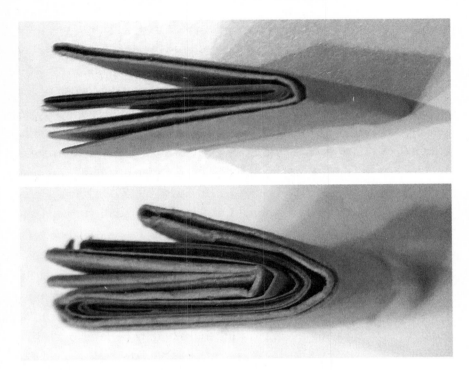

Figure 8.1.

Although a single sheet of paper is very thin, after the seventh folding we would actually have a pile of $2^7 = 128$ layers, which is approximately a finger's breadth in height. Considering that the surface area would have shrunk accordingly, approximately to the dimension of a thumb if we had begun with a letter-sized sheet, it is virtually impossible to accomplish a seventh or clearly an eighth fold, unless we use exceptionally thin paper. While this is indeed surprising and might be a good bar bet, the problem is actually subtler than it seems. First of all, it is the rapid growth rate of the geometric progression 2^n that contradicts our intuition. To illustrate this common misjudgment, consider continuously doubling a pile of standard paper until it reaches the height of the Empire State Building. Starting with a single sheet of paper, it would only require 22 doublings until the pile would surmount the Empire

324 THE MATHEMATICS OF EVERYDAY LIFE

State Building. This example demonstrates how misleading our intuition can be in the context of exponential growth rates.

However, returning to the problem of paper folding, things are a bit more complicated. The number of folds is not limited because of the fact that the force we would have to apply in each additional folding process grows exponentially, eventually exceeding the muscular strength of our fingers. There is a different, purely geometric reason for the existence of a maximum possible number of folds. This was first recognized in 2002 by Britney C. Gallivan (1985–), while she was a high-school student in Pomona, California.[1] She studied the paper-folding problem for extra credit in her mathematics class and showed that there exists a theoretical folding limit for any given piece of paper. This limit cannot be overcome, not even by an arbitrarily powerful folding machine. However, Gallivan also demonstrated that it is possible to fold a sheet of paper considerably more often than just seven or eight times, disproving a common and longstanding belief. How is this possible? Instead of folding in alternate directions, she took a long strip of paper and folded it in a single direction, as we show it in figure 8.2.

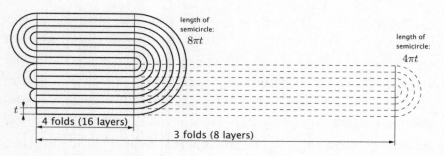

Figure 8.2.

Gallivan used a jumbo roll of industrial toilet paper to fold a single piece of paper twelve times in half. The uncoiled toilet paper measured 4,000 feet in length, and folding it twelve times in half resulted in an approximately one-foot-long packet of paper, consisting of

$2^{12} = 4096$ layers. In 2012, students of St. Mark's school in Massachusetts achieved thirteen folds with a more than 50,000-foot-long strip of industrial toilet paper. To accomplish their task, they spent one Sunday afternoon in the "Infinite Corridor" at MIT, an 825-foot hallway. Gallivan's true achievement was not the setting of a new paper-folding record but the realization that there exists a mathematical folding limit and then deriving an equation for this limit. In figure 8.2, we can see why a sheet of paper can be folded only a certain number of times. In each fold, part of the previous layers will form a rounded edge, and this edge gets bigger with each additional folding. Thus, considering single-direction folding as in figure 8.2, the length of the folded section (where all layers are parallel to each other) will be less than half the length of the previous folded section, since the rounded edge takes up some space. Initially, this curved region is hardly recognizable, but with each additional folding, it will require a larger portion of the volume of the paper (the total volume of the paper does not change). At some point, the length of the folded section will be shorter than what would be needed to create another rounded edge, implying that we cannot double the layers anymore. This means we have reached the folding limit. The thicker the paper, the smaller is the folding limit. (For example, when you fold a hand towel, you will experience this limit after a few folds.) To calculate the folding limit, we only need to find out how much of the paper is bent to a curve in each fold. Denoting the thickness of the paper by t and approximating the curved edges as semicircles of length $n\pi t$, we can compute the length of the paper that gets bent in each fold. Figure 8.3 shows the additional length that is "lost" in each fold:

Number of Fold	Length of Paper That Is Bent	
1	πt	
2	$\pi t + 2\pi t$	
3	$\pi t + 2\pi t + 3\pi t + 4\pi t$	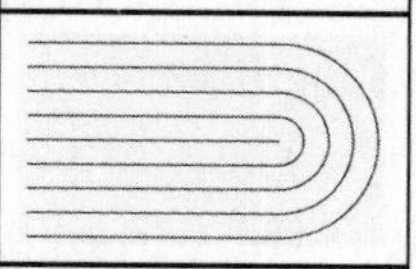
4	$\pi t + 2\pi t + 3\pi t + 4\pi t + 5\pi t + 6\pi t + 7\pi t + 8\pi t$	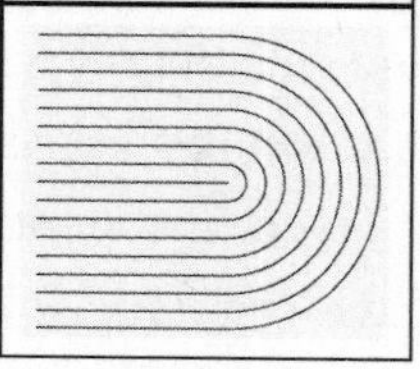

Figure 8.3.

After n folds, the total length of the curved section is therefore

$$\pi t(1 + (1 + 2) + (1 + 2 + 3 + 4) + (1 + 2 + 3 + 4 + 5 + 6 + 7 + 8) + \ldots + (1 + 2 + \ldots + 2^{n-1})).$$

In each of the terms in parentheses, we add up all numbers from 1 to some power of 2, which we may write symbolically as $1 + \ldots + 2^{k-1}$ (k represents the number of the fold). Since the difference between consecutive summands (which are the numbers being added) is constant (which is called an *arithmetic progression*), the arithmetic mean of the numbers to be added is $\frac{1+2^{k-1}}{2}$ (we add the smallest and the largest summand and then halve the result). Multiplying the arithmetic mean by the number of terms being added (2^{k-1}), we obtain a shortcut for the sum, namely, $1 + \ldots + 2^{k-1} = \frac{1+2^{k-1}}{2} \cdot 2^{k-1}$. The latter can also be written as $\frac{2^{k-1}+2^{2k-2}}{2} = \frac{1}{4}2^k + \frac{1}{8}4^k$ and corresponds to the contribution of the k-th fold. Adding up the contributions from all folds, from the first fold to the n-th fold, we thus get

$$\pi t\left(1 + \left(\frac{1}{4}2^2 + \frac{1}{8}4^2\right) + \ldots + \left(\frac{1}{4}2^n + \frac{1}{8}4^n\right)\right).$$

We regroup the terms to be added differently to get the following expression:

$$\pi t\left(1+\frac{1}{4}(2^2+\ldots+2^n)+\frac{1}{8}(4^2+\ldots+4^n)\right).$$

Here the terms to be added form a *geometric progression*, meaning that the ratio between successive terms is constant. We can now use the formula for finding the sum of a geometric progression, which was presented in high school: $S = \frac{a(1-r^n)}{1-r}$, where a is the first term of a progression of n terms, and r is the common ratio between terms. Applying this formula, we find that

$$2^2 + 2^3 + \ldots + 2^n = 2^2 \cdot (2^{n-1} - 1), \text{ and } 4^2 + 4^3 + \ldots + 4^n = 16 \cdot \frac{4^{n-1}-1}{3}.$$

This implies that the total length of all curved sections is $L = \pi t\left(1+\frac{1}{4}\cdot 2^2 \cdot \frac{2^{n-1}-1}{2-1} + \frac{1}{8}\cdot 4^2 \cdot \frac{4^{n-1}-1}{4-1}\right)$, which can be simplified to $L = \pi t\left(2^{n-1} + 2\frac{4^{n-1}-1}{3}\right)$.

This is also the required minimum length of a strip of paper, if we want to fold it in half n times. The following table shows the minimum length for achieving n folds, from $n = 7$ to $n = 13$, if we assume a paper thickness $t = 0.005$ inches (which is a reasonable thickness for standard tissue paper).

Number of Folds n	7	8	9	10	11	12	13
Required Length of Paper Strip	1.2 yd.	4.8 yd.	19.2 yd.	76.5 yd.	306 yd.	1,221 yd.	4,882 yd.

Table 8.1.

An interesting observation is that L increases approximately by a factor of 4 with each additional fold, meaning that you would need four times as much paper as was used in the current record to raise the record by another fold (unless you find a way to use thinner paper). This is also counterintuitive and contributes to the difficulty of tackling the record. Take a piece of paper and fold it in half as often as possible, and you will be surprised how soon you encounter the folding limit!

BUILDING A SKEWED TOWER

When children are playing with stacking blocks, they will sooner or later try to create an overhanging stack where the top block protrudes as far as possible over the bottom block without falling. This is a challenge that comes up quite naturally, and perhaps you even remember your own trial-and-error experiments on this problem. During such attempts, an obvious question will arise: What is the largest possible offset distance that can be achieved between the bottom block and the top block and not fall, if the blocks are only supported by their own weight? If we try to analyze this problem mathematically, we have to set up some rules first. We will begin by assuming that all blocks are identical cuboids (rectangular solids) with uniform density. Then the center of gravity of each block coincides with its geometric center. Moreover, we will assume that each level contains only one block and that each additional block has to be shifted in the same direction, as we show in figure 8.4.

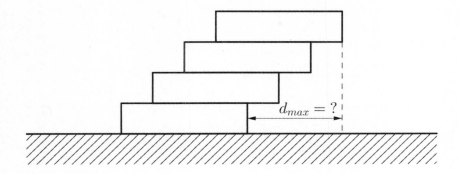

Figure 8.4.

Is it possible to separate the top block from the bottom block by more than the length of one block? Most people's intuition would probably indicate no, and the diagram in figure 8.4 seems to support this belief. Their feeling is that the skewed tower would collapse as soon as the offset between the top and the bottom block exceeds the length of one block. This sounds quite reasonable, but, before exposing whether or not they are right, let us investigate the problem mathematically, starting with the simplest case of only two blocks and then adding more and more blocks. For two blocks, the answer is quite clear, both from an intuitive and from a mathematical viewpoint: The offset between the upper and the lower block must be less than half the length of a block. Imagine slowly slipping the upper block to the right until it topples over. This will happen as soon as its center of gravity passes the edge of the lower block. At that moment, the combined center of gravity (the average position of all the parts of the system) of both blocks together is located "in the middle," that is, at a horizontal distance of $\frac{a}{4}$ from the center of gravity of the lower block, if a is the length of a block (see figure 8.5). The distribution of mass is balanced around the center of gravity. If both blocks were glued together as shown in figure 8.5 and we would apply a force at the combined the center of gravity, the blocks would move in the direction of the force without rotating.

330 THE MATHEMATICS OF EVERYDAY LIFE

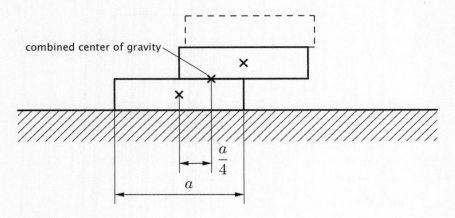

Figure 8.5.

If we would now place another block on top of these two blocks, even the tiniest offset would make the tower collapse, since the third block would then move the combined center of gravity of the two upper blocks at least slightly to the right and thus over the edge of the lowest block. Therefore, if we aim to further increase the overhang by adding more blocks, we should not place the second block's center of gravity directly on the edge of the first block, but start with a smaller overhang. When we then add a third block on the top, we must make sure that the combined center of gravity of the two upper blocks remains above the base of the tower, that is, the lowest block. Placing block D as we show in figure 8.6 would already be "on the edge," since the combined center of gravity of blocks B, C, and D would project exactly onto the edge of block A.

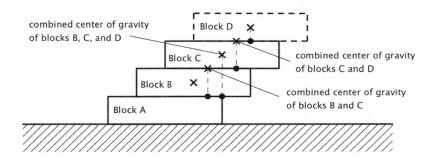

Figure 8.6.

For any block in the tower, the projection of the combined center of gravity of all blocks at a higher level must lie inside the surface of this block. This stability condition must remain satisfied for all blocks in the tower, if we want to add another block on the top. It is very important to recognize that this condition is not about the position of the next block on top in relation to the other blocks, but rather about the position of any block in the stack in relation to all blocks above it. This suggests looking at the problem of producing maximum overhang quite differently. Instead of building a tower from bottom to top, we should think of creating the tower from top to bottom, at least mentally. We start with the topmost block, slide another block underneath it, and continue this procedure. With this way of thinking, it is convenient to number the blocks from top to bottom (block 1 is the topmost block). Let us consider that we already have a tower of n blocks and we want to slide another block underneath it, thereby extending the horizontal distance between the topmost and the bottom block as much as possible. Denoting the horizontal position of the center of gravity of the existing n-block tower by x_n, we place the $(n + 1)$-th block with its right edge at x_n, measured from the right edge of the topmost block. (See figure 8.7.) Since all blocks are identical, the center of gravity of the whole stack is the average position of a block in the stack. It is the arithmetic mean of their centers of gravity. As we have defined it, the average horizontal coordinate of the first n blocks is x_n. If we place the $(n + 1)$-th block with its edge at x_n, its center of gravity will be at $x_n + \frac{a}{2}$. Hence, the combined center of gravity of all $n + 1$ blocks will be at position

$$x_{n+1} = \frac{n \cdot x_n + 1 \cdot \left(x_n + \frac{a}{2}\right)}{n+1} = x_n + \frac{a}{2} \cdot \frac{1}{n+1},$$

since there are n blocks at an average position x_n and one block at position $x_n + \frac{a}{2}$. Thus, adding the $(n + 1)$-th block to the stack allows us to extend the total overhang by $\frac{1}{n+1}$ times half the length of one block. We are now able to answer the question posed earlier: Is it possible to separate the top block from the bottom block by more than the length of one block? Placing the edge of the $(n + 1)$-th block at x_n, we get the maximum overhang in each step. Pursuing this algorithm, the total overhang will exceed the length of one block after only 5 blocks. This is also shown in figure 8.7.

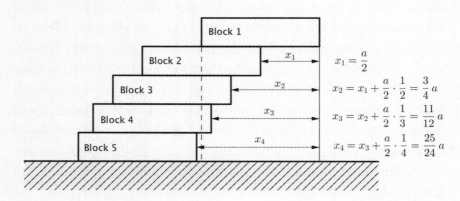

Figure 8.7.

More generally, after n blocks, the total overhang will be

$$\frac{a}{2} \cdot \left(1 + \frac{1}{2} + \frac{1}{3} + \frac{1}{4} + \frac{1}{5} + \frac{1}{6} + \ldots + \frac{1}{n}\right).$$

Here we encounter a partial sum of the famous harmonic series

$$1+\frac{1}{2}+\frac{1}{3}+\frac{1}{4}+\frac{1}{5}+\frac{1}{6}+\ldots$$

Although the sum of all reciprocals from 1 to n grows continuously very slowly with n, it actually has no limit. The harmonic series is divergent, meaning that it "goes to infinity." This implies that there is also no limit for the overhang of a stacked tower of toy blocks or books. However, the number of required blocks increases exponentially with the size of the desired overhang. For an overhang that is twice the length of a single block, we would already need 31 layers; for a triple-block overhang, we would have to stack 227 blocks, always precisely following the algorithm we have described above. Of course, this becomes quite a challenge for a large number of blocks, and it requires a considerable amount of patience.

Calculating the center of gravity is also important for building towers in the real world, especially in modern architecture, where overhanging structures are frequently used. Finding the center of gravity, called the *centroid*, of plane figures such as a rectangle (which is the point of intersection of the diagonals) or of a triangle (which is the point of intersection of the medians) has been part of the typical high-school geometry course. However, finding the center of gravity of an irregular quadrilateral is a bit more complicated. The *centroid* of a quadrilateral is that point on which a quadrilateral of uniform density will balance. This point may be found in the following way. Let M and N be the centroids of $\triangle ABC$ and $\triangle ADC$, respectively. (See figure 8.8.). Let K and L be the centroids of $\triangle ABD$ and $\triangle BCD$, respectively. The point of intersection, G, of MN and KL is the centroid of the quadrilateral $ABCD$.

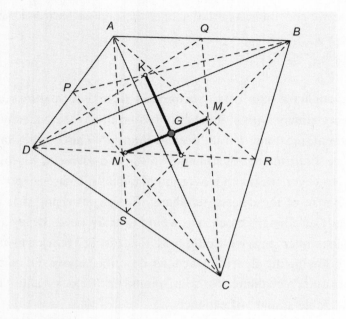

Figure 8.8.

Finding centers of gravity of some figures can be challenging, as you can see from the above illustration. There is much more mathematics involved in the structural analysis of buildings. It is surely fair to say that without the great advances of applied mathematics in the last decades, notably numerical methods and computational mathematics, many architectural masterpieces could not have been realized.

WHISPERING GALLERIES

When we whisper, we usually want to limit the audibility of our speech to nearby listeners. The reason might be to avoid disturbing others in a library, an auditorium, or a place for worship. Whispering is also used to convey secret information without being overheard, for example, during a sales meeting or, to mention a less formal situation, a holiday dinner with weird distant relatives we have never seen before. Quiet, whispered speech can only be recognized very close to the whisperer's mouth. The old but still popular "telephone game" is based on the difficulty of understanding whispered speech and has been played by children around the world for centuries. However, there are places and circumstances in which even the slightest whisper can be heard at a great distance, and, in fact, more distinctively and clearer than a shout.

Rooms with curved and smooth walls can produce a remarkable acoustical phenomenon known as "whispering gallery waves," facilitating that faint sounds can be heard at extraordinary distances. The effect is particularly strong and demonstrative in large rooms with circular, hemispherical, elliptical, or ellipsoidal enclosures. National Statuary Hall in the United States Capitol is a good example. It is a large, two-story, semicircular room beneath a half dome (see fig. 8.9a). From 1807 to 1857, it was the meeting place of the United States House of Representatives.

Figure 8.9a. The National Statuary hall inside the United States Capitol.

There is a myth that the sixth president of the United States, John Quincy Adams (1767–1848), later, as a member of the House of Representatives, took advantage of the hall's acoustics to eavesdrop on other members of the House of Representatives conversing quietly on the opposite side of the room. A floor plaque marks the spot at which Adams's desk on the west side of the hall was located, so tourists can test the acoustics at the other spot. In some spots, a speaker many yards away may be heard more clearly than one closer at hand. Since the half dome, which is in part responsible for the phenomenon, was installed after John Quincy Adams's death, the eavesdropping story was probably created for tourists. However, the curious acoustical properties of Statuary Hall have been documented throughout its history. An August 28, 1892, article in the *New York Times* titled "The Echoes of Statuary Hall" speaks of a "natural telephone" in Statuary Hall: "Away over in the corner you can stand and whisper, while your friends may be at a similar corner on the opposite side of the room and hear every syllable as plainly as though you were standing face to face." The article also states that no one has been able to explain this remarkable phenomenon. Yet a simple geometric explanation already could have been found in the book *Phonurgia Nova* (1673) by the German Jesuit scholar and polymath Athanasius Kircher (1602–1680) (see figure 8.9b).

Figure 8.9b.

Sound waves are mechanical waves of pressure and displacement that propagate through compressible media, such as air or water. When we throw a stone into a pond, ripples form on the water surface and spread out from the entry point as growing concentric circles. Analogously, we may think of sound waves as air-pressure disturbances spreading out from the source as growing concentric spheres. However, unless the source of sound is rotationally symmetric, the intensity of the emitted sound waves will vary with direction. Human speech produces sound waves in the form of concentric spheres whose energy is concentrated in a forward-directed cone. They can be represented by rays that are perpendicular to the growing concentric spheres and lie inside this forward cone. Figure 8.10 shows a diagram drawn by Kircher in which these "sound rays" are reflected from a ceiling in the shape of an ellipsoid of revolution (or spheroid), that is, a surface obtained by rotating an ellipse about one of its principal axes.

Figure 8.10. *Schallreflexion in einem Gewölbe*, authors: Athanasius Kircher and Tobias Nislen, artist: Schultes, Friedrich (Drucker), Deutsche Fotothek, 1684.

Sound waves are reflected from surfaces in just the same way as a ball bounces off a wall. Thus, reflected rays can easily be constructed with a ruler and protractor. If the heads of the speaker and the listener are located in the foci of the ceiling spheroid, then all sound waves going upward to the ceiling will be focused at the listener. The individual reflections can be interpreted as echoes from the ceiling, all heading directly to the listener. Moreover, the sum of the distances to the two focal points is the same for all points on an ellipse. This characterizing property of an ellipse is illustrated in figure 8.11. To draw an ellipse, we may push two pins into a sheet of paper to define the foci of the ellipse and tie a string onto each of the two pins (the string must be significantly longer than the distance between the foci). Then we take a pen and pull the string taut to form a triangle together with the segment $F_1 F_2$. If we now move the pen while keeping the string taut, the tip of the pen will trace out an ellipse.

THE TECHNOLOGY AROUND US   339

This method is called the "gardener's ellipse," since it is also used by gardeners to outline elliptical flower beds.

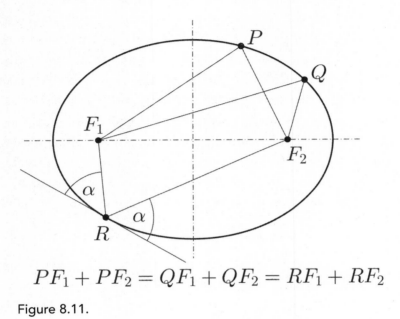

$$PF_1 + PF_2 = QF_1 + QF_2 = RF_1 + RF_2$$

Figure 8.11.

For any two points P and Q on an ellipse with foci F_1 and F_2, the triangles F_1F_2P and F_1F_2Q have the same perimeter. Thus, all sound waves emanating from a focal point and going upward to the elliptical ceiling will have traveled exactly the same total distance when they arrive at the listener, independent of their initial direction. Hence, all echoes will arrive exactly at the same time ("in phase"). This focusing and adding up of sound waves emanating from a focal point of an ellipsoid of revolution makes it possible to hear even the faintest whispers across a large room.

However, the ceiling of National Statuary Hall is hemispherical and not elliptical, so there are not two focal points. As shown in figure 8.12 (*left side*), the center of the hemisphere is the only focal point, but, if the speaker S stands at this focal point and directs his or her speech to the ceiling, then all reflections are focused back to him or her, while the listener L can hardly hear anything. Now suppose that the speaker moves away from the center, as we show on the right side of figure 8.12. Then the reflected rays will concentrate around the point that is symmetric to S with respect to the center of the hemisphere, that is, the point directly opposite to S on the other side of the center and at the same distance from it. It is not a true focal point, but the rays become almost focused there. This "mirror point" would be the ideal position of the listener L. When he comes close to this point, he will suddenly start to hear the speaker talking. Although the almost-focusing effect at the ideal spot gets weaker as the speaker moves farther away from the center (see figure 8.13), there is still a concentration of sound waves around this point. This explains the occurrence of "whispering spots" in National Statuary Hall.

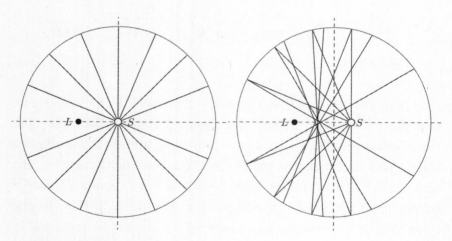

Figure 8.12.

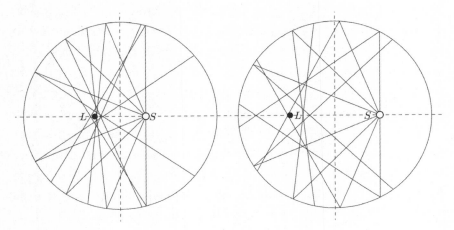

Figure 8.13.

The amplification of sound by the focusing of reflected acoustic waves does not require a room with a curved wall or ceiling. Using two parabolic reflectors positioned opposite to each other, you can set up an acoustic communication channel outdoors (see the section "Looking inside a Flashlight," later, for the focusing property of a parabolic reflector). We show a schematic drawing in figure 8.14. Sound waves emanating from focal point F_1 and directed to reflector 1 will be transmitted to reflector 2 and become focused at its focal point, F_2. Even for distances between the two reflectors well over 200 feet, whispered words can be heard distinctly on the other side.

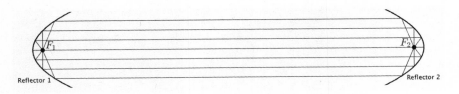

Figure 8.14.

Before the invention of radar, parabolic sound mirrors were used by military air defense to detect approaching aircraft. For demonstrational reasons, acoustic communication channels are still installed in science centers or parks.

A similar effect can be experienced under huge arches with a smooth surface. There is a whispering arch at Grand Central Station in New York City. It is located on the way down to the lower concourse, outside the Grand Central Oyster Bar & Restaurant. The walls and ceiling are made of stone. If there is not too much background noise, two people standing back to back at opposite ends of the underpass can talk to each other by whispering toward the sides of the arch, close to the stone surface. Their conversation cannot be heard anywhere between them, as the sound seems to "creep" along the curved ceiling. This cannot be fully explained on the basis of geometry alone, it also requires some physical understanding of acoustic wave propagation. The location most famous for this phenomenon is the whispering gallery at St. Paul's Cathedral in London (see figure 8.15). It is a circular walkway above the crossing of the nave, 112 feet wide and 99 feet above the floor. A whisper against its curved wall will be audible everywhere along the circumference, even on the other side of the walkway, more than 100 feet away. Although this is already quite astonishing, it is remarkable that a shout is actually less effective than a whisper, which adds to the mysteriousness of the phenomenon.

Figure 8.15. (Image from Wikimedia Creative Commons, author: Femtoquake; licensed under CC BY-SA 3.0 Unported.)

The English physicist and Nobel laureate John William Strutt (1842–1919), now better known as Lord Rayleigh, experimented in St. Paul's Cathedral in the late 1870s to investigate its whispering gallery. He analyzed the problem using the mathematical description of wave propagation and found an explanation for this acoustic curiosity. Sound waves proceeding from a source close to the wall and almost tangential to it will cling to its surface and creep along the circumference of the gallery without losing much of their intensity. Since the involved angles are so slight, the sound travels around the wall by successive reflections along short chords, suffering very little absorption at the hard, smooth wall (see the schematic drawing in figure 8.16). The energy of the traveling sound is concentrated in a narrow belt touching the wall. As we move radially away from the wall, the intensity falls off rapidly. The thickness of the sound belt depends on the frequency of the waves. The higher the frequency, the narrower the belt and the smaller the energy loss along the wall. Since whispers are composed of higher frequencies than normal speech, they are carried farther along the circumference.

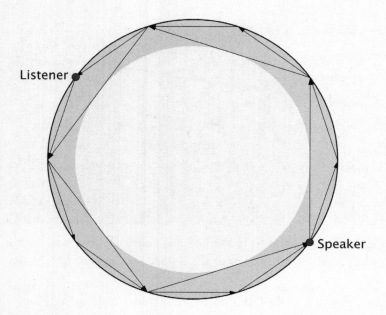

Figure 8.16.

Acoustic waves traveling near the surface of solids are now called *Rayleigh waves* and are also produced by earthquakes. They are used in geophysics to characterize Earth's interior. Thus, Lord Rayleigh's interest in the whispering gallery at St. Paul's Cathedral and his mathematical analysis of surface waves also helped us to gain a better understanding of earthquakes and the composition of the earth. This intriguingly shows the generality of mathematics.

LOOKING INSIDE A FLASHLIGHT

Have you ever wondered why a flashlight with a tiny bulb produces so much light? The luminosity of the bulb or light emitting diode (LED) is actually not the most important factor, but the reflecting surface behind the bulb does the trick. The bulb or LED emits light in random directions, but it is mounted in front of a concave mirror that reflects the light outward. However, the exact shape of the concave mirror is crucial. To get a strong beam of light, we want the flashlight to emit as much light as possible in the same direction. If the emitted light rays are almost parallel, then the flashlight will produce a narrow and bright cone of light. Light whose rays are parallel is called *collimated light*. Lasers produce almost perfectly collimated beams. For a flashlight, the ideal shape for the reflecting surface would be a circular paraboloid, that is, a surface obtained by revolving a parabola around its axis of symmetry (see figure 8.17).

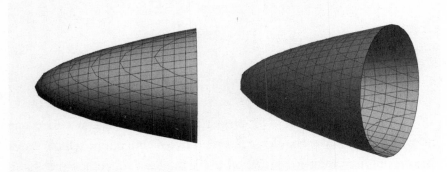

Figure 8.17.

A reflector in the shape of a circular paraboloid is also called a *parabolic reflector*. Before we reveal the geometrical reason for the collimating property of parabolic reflectors, we ought to briefly recall the definition of a parabola.

A parabola is defined by a fixed straight line *d* (the *directrix*), and a fixed point *F* not on this line (the *focus*). The parabola is the set of all points in the plane (which is defined by *d* and *F*) that are equidistant from *d* and *F* (see figure 8.18). This means that a point belongs to the parabola if and only if its distance to the focus *F* is equal to its distance to the directrix *d* (recall that the distance of a point to a line is the length of the perpendicular segment from the line to the point).

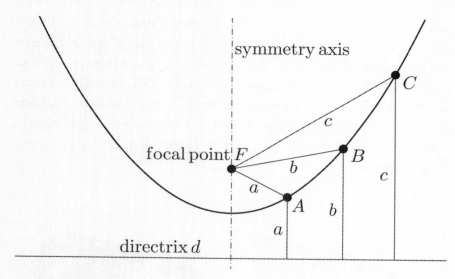

Figure 8.18.

To explore the collimating property of the parabola, we need one more ingredient. We should recall the law of reflection, which states that the angle of incidence is equal to the angle of reflection, measured from the normal. The normal is perpendicular to the surface, or, put another way, perpendicular to the plane tangent to the surface.

We want to show that the reflection of a light ray starting at the focal point *F* of a parabola will be reflected outward and parallel to the symmetry axis. However, since the law of reflection is symmetrical, this is equivalent to showing that the reflection of a light ray coming in parallel to the symmetry axis of the parabola will intersect the focal point F. Let us consider the incident ray being parallel to the parabola's symmetry axis and hitting the parabola in point P as we show it in figure 8.19.

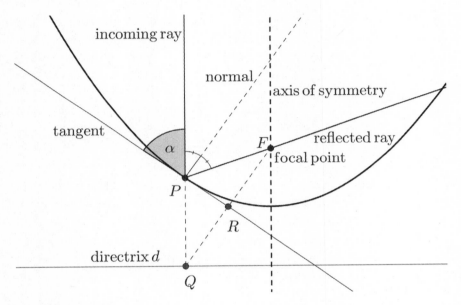

Figure 8.19.

We extend the ray to the directrix and denote the point of intersection by Q. We then draw the tangent to the parabola through P. From our definition of a parabola, we have $PQ = PF$, and PR is the altitude of the isosceles triangle QPF. The angle $\angle FPR$ is therefore congruent to $\angle QPR$. But $\angle QPR = \alpha$, since vertical angles are congruent. Hence, PF indeed represents the direction of the ray reflected from the parabolic mirror. Since P was an arbitrary point on the parabola, this must be true for all points on the parabola, meaning that the reflections of light rays coming in parallel to the symmetry axis will converge at the focal point F. Equivalently, light rays emitted by a bulb or LED mounted at the focal point will be reflected outward, parallel to the symmetry axis, as we show it in figure 8.20. The random light rays emitted by the source are thus directed into a concentrated beam of light.

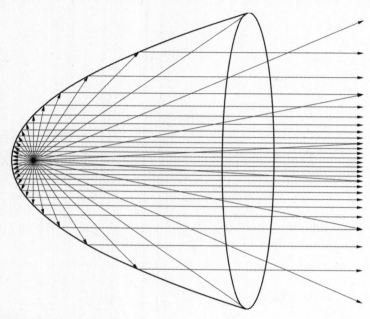

Figure 8.20.

This explains why flashlights produce a strong beam of light even though the light source is relatively weak. Such parabolic reflectors are also used for car headlights.

The reverse principle is used in solar cookers, where a parabolic mirror concentrates sunlight on the focal spot. Figure 8.21 shows a solar reflector with a frame for the pot or pan, mounted in the focal area. There are a hundred ways to grill a burger, but for sure the most ecologically beneficial one, and also not the least fancy one, is to put a grill surface at the focal spot of a solar reflector cooker. Moreover, on the Internet you can find numerous DIY (do-it-yourself) instructions on how to build a solar cooker from cardboard and aluminum foil. It's fun, and it works!

THE TECHNOLOGY AROUND US 349

Figure 8.21. (Image from Wikimedia Commons, author: Nadya Peek; licensed under CC BY 2.0.)

These mathematic principles can go far beyond cooking burgers. According to an ancient tale, Archimedes (ca. 287 BCE–ca. 212 BCE) used parabolic mirrors to burn Roman ships attacking the city of Syracuse in Sicily. This famous episode in the Second Punic War is captured in a wall painting by Giulio Parigi (1571–1635), which we show in figure 8.22. It is probably just a legend, since there is a lot of evidence against the effectiveness of such a weapon. Yet this does not impair the beauty of the principle and its great significance in many applications.

Figure 8.22. (Wall painting from the Stanzino delle Matematiche in the Galleria degli Uffizi [Florence, Italy]. Painted by Giulio Parigi [1571–1635] in the years 1599–1600.)

Satellite dishes represent another application of parabolic reflectors. Like any other form of electromagnetic radiation, signals from satellite radio stations become weaker with increasing distance. Since satellite television signals travel a very long way, approximately 23,000 miles from the emitter to the receiver, they are attenuated considerably when they arrive at the receiver.

Therefore, it is necessary to collect as much of the electromagnetic radiation as possible to ensure interference-free transmission. Radio waves hitting the dish parallel to its axis are reflected to the focal point, where a so-called feed horn converts the signal into electrical currents, which are then transmitted to the satellite receiver. The feed horn is the actual antenna, whereas the much larger dish is only there to collect more radiation and direct it to the feed horn. The parabolic satellite dish acts as geometric amplifier for the radio waves received from the satellite.

The focusing and collimating property of the parabola was probably first proved by the Greek mathematician Diocles (ca. 240 BCE–ca. 180 BCE) in his book *On Burning Mirrors*.[2] The fact that parabolic

THE TECHNOLOGY AROUND US

reflectors are still widely used in many applications today demonstrates both the power and timelessness of mathematical knowledge.

COFFEE WITH CAUSTICS

Perhaps you may have noticed that light reflected from the inside of a ceramic coffee mug sometimes bunches up in a cusped curve on the bottom of the mug. A similar phenomenon occurs when light shines through a drinking glass. The glass will produce curved regions of bright light, with cusps appearing inside the shadow that is cast by the glass. Bright patterns of light can also be observed on the bottom of a swimming pool, when light falls on a rippled water surface. In figure 8.23, we show examples of these phenomena.

Figure 8.23. (Fig. 8.23b from Wikimedia Creative Commons, author: Heiner Otterstedt; licensed under CC BY-SA 3.0 Unported.)

Such concentrations of light are called *caustics*, from the Latin word for "burning," *causticus*, which is itself derived from the Greek καυστός (caustos), meaning "burnt." Caustics are formed when light is reflected or refracted by a curved surface.

When light hits curved glass or a curved surface of water, light rays entering the medium are bent and become focused in certain points or directions. Curved glass or water acts like a magnifying lens. Similarly, when light rays are reflected from the inside of a coffee mug, the curved ceramic wall behaves like a curved mirror that reflects and focuses incident rays. However, the reflected rays are usually not focused in a single point, but instead they get concentrated in a caustic surface, where the density of light will be very high. The caustic curve we can see on the bottom of a coffee mug is the intersection of this surface with the bottom of the mug. To become visible, the concentrated light needs a reflective screen. If there is coffee or tea in the mug, the caustics will be hardly recognizable. However, adding some milk will increase the reflectivity of the liquid surface, thereby making the caustic visible. Again, the curve of bright light we would see on the liquid surface is just the intersection of the caustic surface with the liquid surface.

Although caustics occur under much more general circumstances, let us consider a simplified situation to understand the geometric basis of the formation of a caustic. We will assume that the mug is a circular cylinder and that light shines into the mug from the side (to achieve this, we would have to cut the mug in half, but this shall not concern us now). Moreover, we will assume that the incident light rays are all parallel to each other. This is justified if the source of light is far enough away from the mug. We can then reduce the problem to a two-dimensional picture as we show it in figure 8.24.

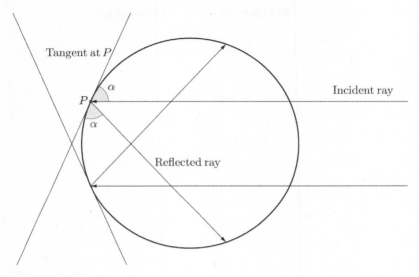

Figure 8.24.

Here, the circle represents the coffee mug and light rays are coming from the right. When a light ray hits the circle from the inside, it will be reflected. Light bounces off circular surfaces at the same angle in which it enters. To find the reflection of a light ray hitting the circle at point P, we first have to draw the tangent to the circle at P. The angle of incidence is equal to the angle of reflection; hence, we can then easily construct the reflected ray. If we do this for a bunch of parallel rays, we get a picture as we show it in figure 8.25.

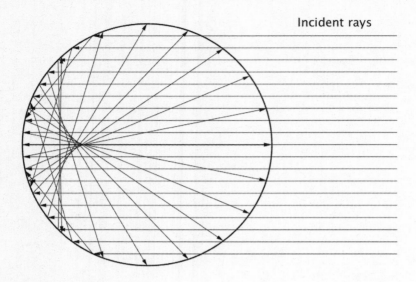

Figure 8.25.

The more rays we draw, the clearer the picture would get. However, we can already anticipate the shape of the emerging caustic. It is the curve to which each of the reflected light rays is tangent. This curve resembles the pattern of light on the bottom of the coffee mug shown in figure 8.23.

Our geometric model explains the formation of caustics, but the caustic emerging in the idealized situation we have shown in figure 8.25 is actually a very special curve. It is a special case of an *epicycloid*, a curve that is produced by tracing a point of a circle rolling without slippage around another circle (recall our earlier discussion about Spirograph toys). Two examples of such curves are shown in figure 8.26, and these two even have their own names. If both circles have the same diameter, we obtain a so-called *cardioid*. Its name is derived from the Greek word for "heart" and refers to its vaguely heart-shaped appearance. A cardioid would be traced by a point on the rim of a coin that is rolled around another, stationary coin. If the outer circle has only half the diameter of the inner circle, then the generated epicycloid is

THE TECHNOLOGY AROUND US 355

called a *nephroid*, which basically means "kidney-shaped." A nephroid would be traced by a point on the rim of a penny rolling around a silver dollar. This is because a silver dollar has exactly twice the diameter of a penny (38.1 mm vs. 19.05 mm).

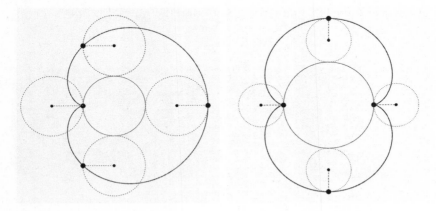

Figure 8.26.

The caustic curve emerging in the geometric construction in figure 8.25 is exactly one-half of a nephroid. A perfect cardioid would arise if the light source were located on the circle. If the light source would be at some finite distance from the circle, we would obtain a caustic curve that is something "in between" a cardioid and a nephroid. Of course, not all of the simplifying assumptions we have made are justified. Light falling into a cup of coffee will typically come from above than from the side, and the shape of the mug will not always be cylindrical. Analyzing these more-realistic situations geometrically would be more complicated, but the appearance of the arising caustic curves would not differ very much from the idealized curves shown in figure 8.26. In particular, they would also display a sharp cusp, which is a typical feature of caustic curves. In figure 8.27 we show three differently shaped mugs that were illuminated by table lamps. In the picture on the left, the caustic curve is a cardioid. A cardioid arises if the cup is roughly a frustum (the remainder when the tip of a cone is cut off, and where the diameter at its bottom is smaller than at its top) and the light shines at an angle equal to the angle of the slant height of the cone (parallel to it). The picture in the middle shows a cylindrical mug and half of a full nephroid at its bottom. In the picture on the right, two light sources were placed symmetrically around the mug to produce a complete nephroid.

Figure 8.27.

The next time you prepare coffee or tea, you can use the time it takes the water to boil for a little experiment. Take a mug with a smooth ceramic finish (white mugs work best) and place it under a lamp or where sunlight can shine into it (you may also use a flashlight). By varying the inclination of the mug in relation to the light source, you will observe various caustic curves resembling those shown in figure 8.27. Using several light sources, you will be able to generate caustics with several cusps. Now that you know how these patterns come about and how they are related to epicycloids, you will look at them with different eyes.

GREEN TRAFFIC LIGHTS ALL THE WAY

If you live in a city with a good public transport system, you don't really need a car. However, as daily traffic jams on the main roads of many cities throughout the world would indicate, not everyone agrees with that.

Of course, even in an urban environment, there can be circumstances in which traveling by car is the only option. On these occasions, traffic signals can be a nuisance to car drivers who are in a hurry. When you are late for an appointment and the traffic lights turn red as you approach each intersection, you might start to believe that these devices are conspiring against you. Yet, sometimes the opposite occurs as well, albeit usually not accompanied by comparably strong but positive emotions. You may have experienced driving on a major thoroughfare and, without any interruption, comfortably passing intersection after intersection with a green light. However, this is not just a matter of luck. Successive traffic lights are often synchronized to ensure a better flow of traffic. In fact, the main purpose of traffic lights is to improve the efficiency of both vehicular and pedestrian traffic. Imagine tight city traffic without any traffic signals! Well, this was the situation in most cities at the beginning of the last century, after which the first automobiles appeared and soon flooded the streets. The Ford Model T, the first affordable automobile, was produced between 1908 and 1927. With 16.5 million units sold, it is still one of the most sold cars of all time. In those early days of motorized private transport, traffic management couldn't keep up with the onslaught of cars. Before the installation of traffic signals in New York City, it could take drivers on Fifth Avenue almost 45 minutes to travel the mile and a quarter between 34th and 59th Streets.[3] The chaotic flow of traffic made it necessary to stop at each intersection, and the lack of organization and control of traffic flows led to almost permanent congestion on workdays. Since automatic traffic signals had not yet been invented, traffic flows at busy intersections had to be controlled by police officers, but the traffic police were not able to deal with the rapidly increasing number of vehicles. The first traffic lights were installed in London in 1868, in reaction to an overflow of horse-drawn traffic near the Houses of Parliament. These signal devices were realized as semaphore arms (an apparatus for signaling with arms or flags) that extended horizontally from a 22-foot-high pillar, with additional red and green gas lamps on top of the pillar for nighttime use (see fig. 8.28). A police constable had to lift and lower the semaphore arms and turn the lamps with a lever so that the appropriate light faced the traffic. In the first two decades of the twentieth century, semaphore traffic signals similar to the one in London were in use throughout the United States. The first three-color traffic light appeared in Detroit in 1920, on one of the busiest intersections in the world. The third light was colored amber, and its purpose was to give police officers a few seconds of time to better coordinate the change from red to green and green to red, respectively. In the 1920s, traffic signals were beginning to be controlled by automatic timers. This was a crucial step in traffic control that led to a dramatic increase of traffic lights. By 1926, New York City had about 100 traffic lights, which were controlled by police officers, and by 1929—merely three years later—the city had more than 3,500 automatic traffic lights. Although the cost of installation was significant, the automated lights enabled the city to reduce the number of police officers in the traffic squad from 6,000 to 500.

Figure 8.28. *Left*: Semaphore traffic signals in London, 1868. *Right*: traffic power in Detroit, 1920.

When the first automatic traffic lights were installed, all lights on an avenue turned green simultaneously. As a foreseeable consequence, cars raced to get through as many green lights as they could before all lights turned red. This produced accidents, and pedestrians were often the ones who suffered as a result of these speeding vehicles. It soon became clear that it is better to coordinate successive traffic signals in such a way that platoons of vehicles can proceed uninterrupted through a series of green lights. This not only improved safety but also ensured a better flow of traffic and minimized gas consumption and pollutant emissions. With the idea of optimizing traffic-signal coordination, mathematics enters into the game. Finding the optimal timing pattern of traffic lights at successive intersections is essentially a mathematical problem. For simplicity, we will assume that we have a sequence of intersections where the distance d between two successive intersections is always the same. Each intersection is equipped with identical traffic lights, in particular, the cycle duration T and the phases within a cycle (green, yellow, red) are the same for all intersections. Now assume that a car gets through the first intersection just after the lights switched to green. Depending on its average speed v, the car will need a time $\Delta t = \frac{d}{v}$ to get to the next intersection. Therefore, if we want the green phase at the second intersection to start just as the car arrives, we have to shift the traffic signal cycle at the second intersection exactly by a time lag $\Delta t = \frac{d}{v}$. In fact, if we assume that the average speed of the car is constant, the traffic signal cycle must always be shifted by this amount, when we go from one intersection to the next. In a snapshot showing the phases of successive traffic lights at a certain moment in time, we will thus get a periodic pattern as shown in figure 8.29. We may think of this pattern as a wave whose wavelength is equal to the (spatial) distance between two intersections whose traffic lights are (always) in the same phase. For instance, in figure 8.29 the third and the tenth traffic light are completely synchronized, hence the wavelength is equal to the distance between the second and the tenth intersection.

360   THE MATHEMATICS OF EVERYDAY LIFE

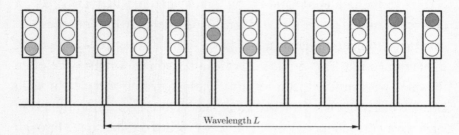

Figure 8.29.

In figure 8.29, the first two traffic lights on the left are in the green phase. The third traffic light from the left is in the red phase, but it will soon switch to green. The sixth traffic light is in the yellow phase, meaning that it will soon switch to red. Recall that the duration of a cycle is T. This means that a snapshot taken a time T later will look exactly the same as the one shown in figure 8.29. However, if we take another snapshot after a time interval that is only a fraction of T, the picture will look different. Figure 8.30 shows snapshots at different times within one cycle, and we can see that the pattern we called a wave actually travels from left to right as time proceeds.

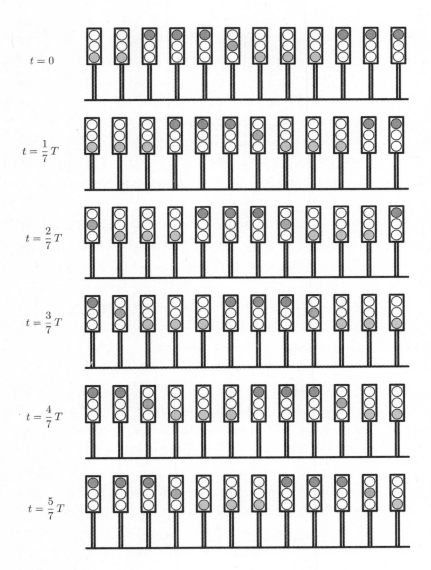

Figure 8.30.

In fact, this is the true reason why traffic lights coordinated in this way are sometimes also called a "green wave." Waves are essentially patterns that propagate, and so does the pattern of traffic signals in figure 8.30. As time passes, it propagates from left to right. More precisely, as can also be inferred from figure 8.30, the wave travels exactly one wavelength L during each signal cycle of duration T. The speed of the wave is therefore $\frac{L}{T}$, and for a car traveling at exactly the same speed, green lights will cascade one after the other. The speed of the traffic signal wave can be adjusted by means of the time lag between the signal cycles at two successive intersections. The smaller the time lag, the faster the wave will propagate. If we want to allow platoons of vehicles to proceed through a series of green lights, we have to adjust the time lag Δt to the expected average vehicle speed v and the intersection spacing d. The condition $\Delta t = \frac{d}{v}$ ensures that cars getting through a particular intersection during the green phase will also get a green light at the following intersection, provided they maintain the optimum speed v, which is usually chosen to be slightly below the speed limit. Moving with constant speed $v = \frac{L}{T} = \frac{d}{\Delta t}$ means traveling along with the wave, quite similar to surfing on water waves. Cars moving slower or faster than v will run out of phase with the wave, and, therefore, will encounter a red light sooner or later.

The first coordination of traffic lights implementing this scheme was realized in Salt Lake City in 1917, where the manually controlled signal cycles at six successive intersections were coordinated to allow for a smooth flow of traffic. In modern coordinated signal systems, it is possible for drivers to travel long distances before encountering a red light, typically at times when there is less traffic, since even a very short traffic holdup will decrease the average speed and bring you out of phase with the traffic signal wave. However, any coordination is futile during traffic congestion. Moreover, traffic signals can also be coordinated "against drivers," that is to intentionally slow down traffic; for instance, to prevent high volumes of traffic at bottlenecks and reduce the risk of congestion.

The principle of traffic signal coordination can be illustrated very nicely by using a representation also known as a space-time diagram (see fig. 8.31). Here the time axis is drawn upward and the horizontal line represents the spatial distance from the origin. The curves drawn in a space-time diagram are referred to as "world-lines." For example, the world-line of a car at rest is represented by a vertical line in the diagram, parallel to the time axis, since the spatial position of the car does not change with time. The world-line of a car moving at constant speed $v > 0$ is a straight line with slope $\frac{\Delta t}{\Delta x}=\frac{1}{v}$; that is, the higher the speed, the more the world-line will be tilted toward the horizontal direction. A horizontal world-line would mean infinite speed, which is not possible by the laws of physics, since nothing can travel faster than the speed of light. The shaded regions in figure 8.31 represent the red phases of the corresponding intersections, drawn schematically below the horizontal axis. Thus, the world-line of a car that passes through an intersection during the green phase will run between two consecutive shaded rectangles at the position of this intersection. If the world-line of a car runs into a shaded rectangle, the car must stop because it reaches the intersection during the red phase. While the driver waits for the lights to turn green again, the world-line of the car goes on vertically, meaning that time passes, but the car is at rest. As the diagram demonstrates, the fixed time lag between the signal cycles of successive interactions corresponds to a vertical shift of the shaded rectangles from intersection to intersection, allowing cars with the right speed to encounter a cascade of green lights and pass smoothly through a series of intersections.

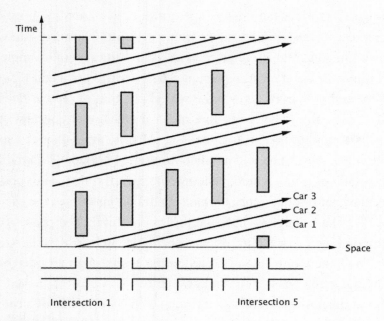

Figure 8.31.

The diagram in figure 8.31 also shows that the coordination of traffic signals for one direction will in general have the opposite effect on traffic flowing in the other direction. The world-line of a car moving at a normal speed from right to left will inevitably run into a red-light region at every second intersection (see fig. 8.31). Hence, traffic signal coordination does not work for both directions on two-way streets. However, many cities in the United States use traffic signal coordination also on two-way streets, operating in the direction more heavily traveled. This direction may change during the day. For example, many commuters travel from the suburbs to the city in the morning, but in the opposite direction after work. Therefore, the lights are timed to accommodate the rush-hour traffic's direction. For example, in New York City there are a number of such two-way streets that accommodate rush-hour traffic, such as Riverside Drive and Central Park West in Manhattan.

We have only touched upon some mathematical aspects of traffic signal coordination, which is much more complex in reality. Signal cycles may be different at different intersections and vary during the day, depending on cross traffic. In addition, waiting times for pedestrians to finish crossing the street have to be considered, and many other factors have to be taken into account as well. Nowadays, computerized control and coordination systems enable traffic flow organization to react almost instantaneously to critical traffic situations and automatically adjust traffic signal cycles to minimize delay. All of these systems rely on mathematics, and finding the optimal solution basically means to solve a complicated system of equations, much more complicated than the simple example just discussed.

SAFETY IN NUMBERS

When you were a child, you may have exchanged secret messages with a friend by using some sort of secret code. One primitive example of a code is mirror writing. The message will look very strange at first sight, and an uninitiated reader may conclude that it must have been written in a foreign language. However, if writing in a reverse direction is perfectly executed, then the message will appear completely normal when it is reflected in a mirror. Interestingly, Leonardo da Vinci (1452–1519) wrote most of his personal notes in mirror style, for example, the notes on his famous Vitruvian Man, part of which are highlighted in figure 8.32.

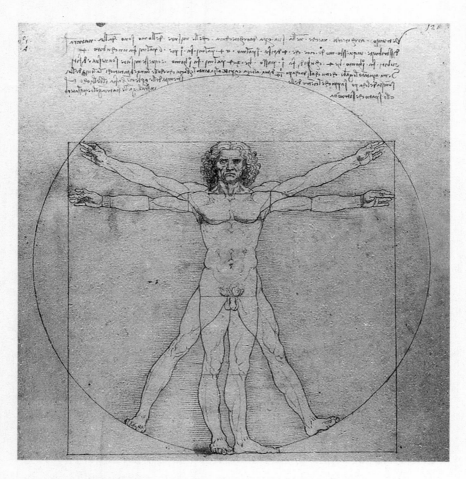

Figure 8.32.

Although there exist various speculations, it is not really known why he wrote this way. Mirror writing is not a very effective way to prevent somebody else from reading a message, yet it will at least considerably slow down one's reading, unless, of course, a mirror can be used to reflect the message. Some other reasons have been suggested, for instance, the fact that Leonardo probably was left-handed might have motivated him to write from right to left to avoid smudging the ink with his hand.

A famous historical person who used a code to protect secret messages was Julius Caesar (100 BCE–44 BCE). The technique he used is now known as "Caesar's code" or "Caesar's cipher." A cipher is an algorithm to convert information into a code, which is where the word *decipher* comes from. We will explain the Caesar cipher in a moment, but let us first introduce some terms used in cryptography, which is the study and practice of secure communication. The word *cryptography* stems from the Greek words *kryptós* for "secret" and *graphein* for "writing"; thus, it literally translates to "secret writing." The information in its original form is known as *plaintext*. Applying the cipher algorithm, it can be *encrypted* or encoded, which is to convert written material into a form that can only be accessed by authorized parties who are familiar with the code. This encrypted message is also known as *ciphertext*. An authorized person can "undo" the encryption, a process called *decryption*, with the help of a so-called *key*, which specifies the transformation of plaintext into ciphertext.

So how does the Caesar cipher work? Each letter in the plaintext is replaced by a letter some fixed number of positions down the alphabet.[4] The Roman historian Suetonius (ca. 69–after 122) wrote about Caesar as follows in *The Twelve Caesars*:

> If he had anything confidential to say, he wrote it in cipher, that is, by so changing the order of the letters of the alphabet that not a word could be made out. If anyone wishes to decipher these, and get at their meaning, he must substitute the fourth letter of the alphabet, namely D, for A, and so with the others.

368 THE MATHEMATICS OF EVERYDAY LIFE

This encryption scheme is shown in figure 8.33. Each letter in the plaintext is shifted three positions further down the alphabet.

A	B	C	D	E	F	G	H	I	J	K	L	M	N	O	P	Q	R	S	T	U	V	W	X	Y	Z
D	E	F	G	H	I	J	K	L	M	N	O	P	Q	R	S	T	U	V	W	X	Y	Z	A	B	C

Figure 8.33.

To illustrate how the algorithm works, let us encrypt the message "*rerum omnium magister usus*," a quote by Julius Caesar, which translates to "experience is the teacher of all things":

RERUM OMNIUM MAGISTER USUS = UHUXP RPQLXP PDJLVWHU XVXV

We look up the first letter of the plaintext, R, in the first row of figure 8.33, and replace it by the letter directly below, U. The same procedure is applied to the rest of the plaintext, resulting in the ciphertext "uhuxp rpqlxp pdjlvwhu xvxv." To decrypt or decipher this message, you have to apply the reverse operation, that is, shift each letter three positions up the alphabet. If you know the essential mechanism behind this decryption technique, it is actually not too difficult to "crack the code," since there are only twenty-five different possibilities of shifting each letter a fixed number of positions in the alphabet. The strategy is simple: Just try all possible shifts and you will eventually recover the plaintext. If the ciphertext is long enough, the frequency of the occurrence of certain letters may provide an "educated guess" for the shift value. For instance, in the English language, the letter E occurs quite frequently, while Q or Z may give you a hard time in a Scrabble game. If you decrypt a (sufficiently long) message written in English by using Caesar's cipher, then there is a good chance that the most frequent letter in the ciphertext will be the one that replaced every E in the plaintext. This tells you the value of the shift, which is basically the "key" necessary to decrypt the message.

THE TECHNOLOGY AROUND US 369

A more-sophisticated cipher was first described by Giovan Battista Bellaso (1505–?), but it was reinvented and refined by Blaise de Vigenère (1523–1596), and therefore it is known today as the "Vigenère cipher." Although the basic principle is again shifting letters up or down the alphabet, the Vigenère cipher is a much safer way of protecting messages than Caesar's cipher. To encrypt a message, you use a so-called Vigenère table, which is simply a square of shifted alphabets, as shown in figure 8.34, and, as the second ingredient, a secret keyword.

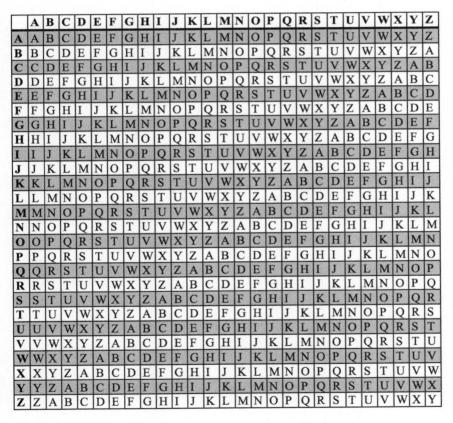

Figure 8.34.

The essential idea is to pair each letter in the plaintext with a letter in the keyword to generate the ciphertext. If the message is longer than the keyword, then the keyword is repeated again and again. Let us demonstrate the scheme in an example. Suppose our keyword is "peanut" and the plaintext of our message is "The cake is in the drawer." The first row of figure 8.35 contains the plaintext, and the second row contains the repeated keyword. Now we use the Vigenère table to encrypt the message.

T	H	E	C	A	K	E	I	S	I	N	T	H	E	D	R	A	W	E	R
P	E	A	N	U	T	P	E	A	N	U	T	P	E	A	N	U	T	P	E
I	L	E	P	U	D	T	M	S	V	H	M	W	I	D	E	U	P	T	V

Figure 8.35.

Any pair of two letters can be thought of as a pair of coordinates on the Vigenère table. For instance, interpreting the letters in the plaintext as "horizontal" coordinates and the letters in the keyword as the "vertical" coordinates, the first pair of letters, consisting of one letter from the plaintext and one from the keyword, (T, P), yields the letter I, since this is the letter at the intersection cell of column T and row P in the Vigenère table (see figure 8.34). The next pair, (H, E), gives L, and so on (see the third row in figure 8.35). In this way, we obtain a ciphertext, which is much harder to decrypt than the simple Caesar's cipher, as we are actually using several Caesar's ciphers with different shift values. Moreover, the shift values depend on both the plaintext and the keyword. To decrypt and encrypt a message, the keyword must be known. Hence, before two people, say, Lisa and David, can communicate through encrypted messages, they have to exchange a keyword, or, more generally, a "key string" of letters. This might as well be the sequence of letters in a certain book they both possess, making it very hard for a third person (an eavesdropper) to decrypt their messages. But before any secure communication between two parties can start, they must somehow exchange keys. However, if a third party finds out the key (by spying, for instance), the communication is no longer secure. That's why the key has to be changed from time to time (similar to how it is recommended that you renew your passwords for online services at regular intervals). It is not very practical if Lisa and David have to meet in person every time they want to renew their key, and, even then, there is always a risk that an eavesdropper is watching or listening in on the exchange. Exchanging the key is actually the crucial point in setting up a secure communication channel. No matter how complex the decryption algorithm is, the communication between Lisa and David will only be secure as long as the key is not known by anyone else. But how can they be sure that their key has not yet been discovered by a third party? Well, it has long been thought that they can never be sure, and, thus, any cryptographic system can be broken, because there is no way to exchange a key without leaving at least a tiny loophole for potential eavesdroppers. Surprisingly, there exist two wonderful solutions to the problem of secure key exchange: one is based on an ingenious mathematical idea, and one is provided by quantum physics. We will not delve into the latter, but we will try to explain the brilliant mathematical idea underlying what is now known as the "Diffie-Hellman-Merkle key exchange." In short, the method allows two parties to jointly establish a shared secret key over an insecure channel. There is no need to use secure communication, or institute any other measures of security in the key-exchange process, since the key itself is not actually exchanged, at least not in the usual sense. Both parties create their joint key "at home," and the only information they need to exchange for this purpose can even be transmitted over a public channel, or, for example, even on a public website. So, how does this work?

Let us first consider a simple, but insecure, method of how Lisa and David can create a joint key without having to transmit the key itself. Lisa and David agree on an arbitrary integer, g. Then Lisa chooses an integer a, computes g^a, and sends this number to David. Then David chooses an integer, b, computes g^b, and sends the result to Lisa, who then takes the a-th power of the number she received, obtaining $(g^b)^a = g^{b \cdot a}$. Continuing along, David, on the other hand, received the number g^a from Lisa, which he raises to the b-th power, thereby getting $(g^a)^b = g^{a \cdot b} = g^{b \cdot a}$. Therefore, both Lisa and David end up with the same number: $g^{b \cdot a}$, which was created in a special way from the numbers they chose independently of each other. This number could serve as a joint encryption key, but, unfortunately someone could intercept their communication and get to know g, g^a, and g^b. It would then only be a matter of trial and error to manipulate and combine these numbers in the right way to obtain the key $g^{a \cdot b}$. In this process, however, an eavesdropper would have to compute a from g^a, or b from g^b. This operation, which is the inverse of exponentiation, is called "the logarithm to the base g." More precisely, a number a is called the logarithm to base g of a number r, if $g^a = r$. Here, g and r are assumed to be positive real numbers, and a can be any real number. For example, the logarithm to the base 2 of 8 is 3, since $2^3 = 8$, and the logarithm to the base 3 of 81 is 4, since $3^4 = 81$. If the considered numbers are larger, one may not be able to find out the logarithm by using paper and pencil, but it can still be determined without effort under the assistance of an electronic calculator. For very large numbers, say, with hundreds of digits, one may need a really powerful computer, but it can still be done in reasonable time with enough computational power.

Essentially, the ingenious idea underlying the Diffie-Hellman-Merkle method is to modify the generation of the joint key $g^{a \cdot b}$ via exponentiation in such a way that the inverse operation of determining a from g^a or b from g^b becomes significantly harder in the sense that finding the solution would consume an immense amount of time, no matter how powerful your calculating machine may be. The safety of the Diffie-Hellman-Merkle key exchange relies on the fact that even the fastest computers in the world would need more time than the age of the universe for this task.

To reveal the trick that is used in this method, we have to introduce another mathematical operation, which is called the *modulo operation*. Basically, the modulo operation finds the remainder after division of one integer by another. For example, 8 mod 2 = 0 (which is read as "8 is congruent to 0 modulo 2" or, shorter, "8 modulo 2 is 0"), because if we divide 8 by 2, the remainder is 0. Analogously, we have 20 mod 2 = 0, but if we divide an odd number by 2, then the remainder will be 1, for example 5 mod 2 = 1. More generally, all even numbers are congruent to 0 under the modulo-2 operation, and all odd numbers are congruent to 1. Calculating modulo 3, we have three different possible remainders after division by 3, so all numbers will fall in one of three "classes" determined by the remainder after division by 3. To visualize the modulo operation, we may think of a polygon with k vertices, labeled counterclockwise from 0 to $k - 1$.

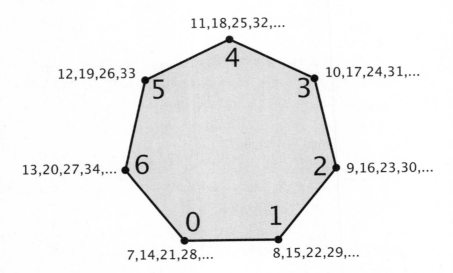

Figure 8.36.

Then, to determine n mod k, we start at vertex 0 and move exactly n steps counterclockwise, hopping from vertex to vertex. The number of the vertex at which we finally end up, is n mod k. In figure 8.36, this procedure is shown for the case $k = 7$, using a 7-gon. All natural numbers are congruent modulo 7 to one of the numbers 0,1,2,3,4,5, and 6. For example, 10 mod 7 = 3, since 10 steps will lead us from vertex 0 to vertex 3. In the Diffie-Hellman-Merkle method, the modulo operation is used to give the joint key creation as described above a decisive twist. First, Lisa and David agree on some prime number p and a number g between 1 and $p - 1$, such that g^k mod p can take on any value from 1 to $p - 1$, if k runs through all natural numbers. As an example, let us consider $p = 7$ and $g = 3$ (see figure 8.37).

k	0	1	2	3	4	5
3^k	1	3	9	27	81	243
3^k mod 7	1	3	2	6	4	5

Figure 8.37.

Next, Lisa chooses a positive integer a and computes g^a mod p. Let us assume she chooses $a = 4$. Then, using the table in figure 8.37 and the 7-gon in figure 8.36, she obtains the following:

g^a mod $p = 3^4$ mod 7 = 4 mod 7

and sends this number, 4, to David. Analogously, David chooses a number b, say, $b = 5$, and computes:

g^b mod p = 3^5 mod 7 = 5 mod 7.

Here we may use that $x \cdot y$ mod $p = (x$ mod $p)$ $(y$ mod $p)$, which is clear since replacing a factor x of a composite number by $(x$ mod $p)$ means that we subtract a multiple of p from the number, which doesn't change the result modulo p. Now David sends his result, 5, to Lisa. Just as described earlier, Lisa computes $(g^b \bmod p)^a \bmod p = (5^4 \bmod 7) = 2$, and David computes $(g^a \bmod p)^b \bmod p = (4^5 \bmod 7) = 2$. Again, both parties will obtain the same result, since $(g^b)^a = (g^a)^b$. Thus, they have created their joint decryption key, just as in the original setting. However, let's switch to the role of an eavesdropper, say, Evan, where we see that the situation is now different. Listening in on the exchange of numbers, he finds out that $p = 7$, $g = 3$, $g^a = 4 \bmod 7$, and $g^b = 5 \bmod 7$. (Lisa and David may even communicate this information publicly.) To determine the secret key $(g^b)^a \bmod p$, Evan has to determine a, meaning that he has to take the logarithm of g^a to base g, modulo p. This operation is called the *discrete logarithm*. But while the "classical" logarithm to base g can be computed very quickly even if g^a has hundreds of digits, there exists no efficient method to compute the discrete logarithm. Essentially, one has to compute $g^k \bmod p$ for all numbers k between 1 and $p - 1$ and compare the result with g^a. For instance, our eavesdropper, Evan, has to compute $3^k \bmod 7$ for all values of k from 1 to $7 - 1 = 6$. As soon as he obtains the result 4 (which is $g^a \bmod 7$ in our example), he can stop. This is the case for $k = 4$ (see figure 8.37). He then knows the number a that Lisa had chosen. With this knowledge, he can calculate Lisa's and David's secret key $g^{ab} \bmod 7$ (which was 2), since he already knows that $g^b = 5 \bmod 7$:

$$g^{ab} \bmod 7 = (g^b)^a \bmod 7 = 5^4 \bmod 7 = 5^2 \cdot 5^2 \bmod 7 = 16 \bmod 7 = 2,$$

which is exactly the key Lisa and David had established. Looking at the table in figure 8.37, we see that Evan had to compute the first four powers of 3 until he got the desired result 4 mod 7. Of course, this can be done rather quickly, but here we just presented a simple example to demonstrate the basic principle behind the Diffie-Hellman-Merkle key exchange. The prime numbers p typically used in real applications have approximately 300 digits, meaning that $p \approx 10^{300}$ and, therefore, an eavesdropper would have to compute and test 10^{300} numbers individually. Notice that the number 10^{300} is a 1 with 300 zeros attached! Considering that the number of atoms in the observable universe[5] is only about 10^{80}, we would have to replace each atom in the universe by a whole universe the size of our own universe, to obtain $10^{80} \cdot 10^{80} = 10^{160}$ atoms, and then repeat this process once more to finally gain a number of atoms which is larger than 10^{300}. It is obvious that even for modern supercomputers, computing the discrete logarithm of such incredibly large numbers is a task that cannot be achieved in a reasonable amount of time.

The Diffie-Hellman-Merkle key exchange and similar methods based on the virtual impossibility of solving certain mathematical problems are used for secure data transfer in a variety of internet services. Secure online payment systems, which we frequently use, would not be possible without the mathematical concepts underlying modern cryptography. Here we see how mathematics can help to keep us secure!

THE ISBN SYSTEM

On the back cover of this book, you find a bar code representing the ISBN of this book, that is, its International Standard Book Number. The ISBN system was developed by the International Organization for Standardization (ISO) and was published in 1970 as the international standard ISO 2108. It was based on the Standard Book Number (SBN) system, which had already been in use in the United Kingdom since 1967.[6] Before the ISBN system was introduced, one had to go through the title, publisher, and year of publication to identify a particular edition of a book. The ISBN system considerably simplified the search for a specific printing of a book and facilitated the use of electronic order transmission systems, which became widespread in book trade in the 1970s. A worldwide numbering scheme for books was also essential for bibliographic records in electronic form, paving the way for electronic databases in libraries and bestseller lists. Each edition of a book has a specific ISBN code. For example, an ebook, a paperback, and a hardcover edition of the same book would each have a separate ISBN. The ISBN codes assigned in 2007 or later are 13 digits long, and they are 10 digits long if they were assigned before 2007. It is a "structured" number—meaning that different parts of the number have different meanings (similar to ZIP codes). The parts of an ISBN are usually separated by hyphens. Any 10-digit ISBN can be converted into a 13-digit ISBN. In fact, the 9 digits before the last one are identical in both variants. Apart from a different last digit, the ISBN-13 only has an additional 3-digit prefix to make the ISBN code compatible with the 13-digit international article number system EAN-13, which has already been used for products other than books. You will find EAN barcodes on all sorts of products in your local supermarket. EAN-13 codes start with a 3-digit country prefix, but all ISBN-13 codes start with "978," regardless of the country of origin. The universal country code "978" is, therefore, informally known as "Bookland." For example, the first edition of this book (hardcover version) has the ISBN 978-1-63388-387-1. The prefix "978" is followed by a "registration group element" indicating the country, geographical area, or language ("1" represents English), which is then followed by a "registrant element" exactly identifying the publishing house and its address, and then a "publication element" identifying the specific publication of that registrant. The last digit is a so-called check digit, and this is where mathematics enters into the game.

A check digit, usually appended at the end of a number code, is calculated from the previous digits in the code. It does not contain any additional information. Its role is merely to detect errors in the preceding digits. Errors mostly occur when numbers are typed or written down on paper and put into a computer manually. Check digits enable computers to detect such errors immediately. Imagine you have found a several-years-old special edition of a certain book in the town library, and you would like to purchase it to get your own copy. You write down its ISBN and go to a bookstore to find out whether it is still available. Unfortunately, you got one of the digits wrong, which you didn't notice. However, as soon as your bookseller enters the number written on your note into his ISBN search engine, he will tell you that this is not a valid ISBN and that you must have made a mistake. The two most common errors in handling an ISBN are a single altered digit or the transposition of two adjacent digits. The check digit methods of ISBN-10 and ISBN-13 ensure that a wrong number arising from the alteration of a single digit of a valid number will always be detected. In addition, the ISBN-10 check digit method also allows us to identify any wrong number arising from the transposition of two adjacent digits, while the ISBN-13 check digit does not catch all errors of this kind.

How does the check digit method work? The check digit appended at the end of an identifying number (which could be an ISBN, a bank account number, or an order or customer number), is calculated from the preceding numbers by a mathematical formula. This formula is chosen in a way that the obtained result, that is, the check digit, will be definitely different if one of the preceding digits is altered. Of course, this would, for example, be true for the sum of all the preceding digits. However, since the check digit is only a single digit, we cannot simply take the sum of all the preceding digits. The formula uses the modulo operation, which we introduced in the previous section, "Safety in Numbers." So how does this work?

Let us first look at the calculation of the ISBN-10 check digit. Consider the nine-digit number 1-63388-387, obtained from the ISBN-13 of this book by deleting the first three digits (the "prefix element") as well as the last digit. If this remaining number were to represent the first nine digits of an ISBN-10, then its tenth digit, the check digit, would be calculated as follows: First, multiply the first digit by 10, the second digit by 9, the third digit by 8, ..., the ninth digit by 2, and add up all these numbers:

$$10 \cdot 1 + 9 \cdot 6 + 8 \cdot 3 + 7 \cdot 3 + 6 \cdot 8 + 5 \cdot 8 + 4 \cdot 3 + 3 \cdot 8 + 2 \cdot 7 = 247$$

This is called a *weighted sum* (the descending sequence 10, 9, 8, . . ., 2 are the weights of the digits). The check digit is the smallest number you have to add to this weighted sum to get a multiple of 11: $247 + 6 = 23 \cdot 11$, hence the check digit is 6. With the help of the modulo operation, we can write this as $11 - 247 \mod 11 = 6$, that is, we get the check digit by subtracting the weighted sum modulo 11 from 11. Note that the result must always be a number from 0 to 10 (since we are calculating up to multiples of 11). It cannot be larger than 10, but what would the check digit be if the result would have been 10? In this case, you would write "X" in the check digit place, which could be interpreted as the roman symbol for 10 (see the section on roman numerals earlier in this book). Thus, the check digit in the ISBN-10 system can be any symbol from the set {0, 1, 2, 3, 4, 5, 6, 7, 8, 9, X}, where "X" represents "10." The correct ISBN-10 code for this book is therefore 1-63388-387-6, and the last digit is the ISBN-10 check digit we have just calculated. Assume now that we alter one of the other digits, writing p instead of q. Then the weighted sum of the first nine digits would change by an amount $(11 - k)(p - q)$, where k is the position of the altered digit. Is it possible to obtain the same value for the check digit? The check digit would only stay the same if $(11 - k)(p - q)$ were a multiple of 11, meaning that $k(p - q)$ would have to be a multiple of 11 or that $k(p - q) = 11 \cdot n$ for some integer n. But this is impossible, since 11 is a prime number and cannot be decomposed into smaller factors. The equality $k(p - q) = 11 \cdot n$ would imply that 11 is a divisor of k or of $p - q$. But k ranges from 0 to 9 and $p - q$ ranges from -9 to 9. Therefore, neither of them is divisible by 11. Consequently, the value of the check digit must change as soon as we alter one of the first 9 digits.

What happens if we switch two adjacent digits within the first 9 digits? This is called a "transposition error." Assume that the correct number has digit p at position k, and digit q at position $k + 1$. Exchanging p and q would change the value of the k-th term in the weighted sum by $(11 - k)(q - p)$ and the value of the $k + 1$-th term by $(11 - (k + 1))(p - q)$. In total, the change would be: $(11 - k)(q - p) - (11 - (k + 1))(q - p)$.

Again, we may ask ourselves, "Is it possible that this expression is a multiple of 11"? The answer is clearly no, because the total change is actually equal to $q - p$ (we encourage you to verify this), which ranges from -9 to 9 and therefore cannot be divisible by 11.

Our considerations have shown that if we compare the last digit of an ISBN-10 with the check digit we have calculated from its first 9 digits, we are able to detect all single transposition errors and all single-digit alterations.

Let us also take a look at the ISBN-13 now. Here the check digit is calculated in a slightly different way. Again, one computes a weighted sum from the first 12 digits, but one alternately assigns the weights 1 and 3 to the digits, starting with 1. Let us check whether the check digit of the ISBN-13 of this book is correct. The first 12 digits are 978-1-63388-387, so we get:

$$1 \cdot 9 + 3 \cdot 7 + 1 \cdot 8 + 3 \cdot 1 + 1 \cdot 6 + 3 \cdot 3 + 1 \cdot 3 + 3 \cdot 8 + 1 \cdot 8 + 3 \cdot 3 + 1 \cdot 8 + 3 \cdot 7 = 129.$$

The check digit of the ISBN-13 system is the smallest number you have to add to this sum to get a multiple of 10: $129 + 1 = 13 \cdot 10$, hence the check digit is indeed 1. Altering one of the digits from q to p will change the value of the weighted sum by $(p-q)$ or $3(p-q)$. Clearly, $p - q$ cannot be a multiple of 10. Since the expression $3(p - q)$ must be some multiple of 3 between -27 and 27, it cannot be a multiple of 10 either. Since any alteration of a single digit will render the value of the total sum by an amount that is not divisible by 10, the value of the check digit will change as well. Thus the ISBN-13 check digit method ensures that single-digit alterations can be detected. However, if two adjacent digits are switched, the weighted sum in the ISBN-13 system will change by $\pm((p-q) + 3(q-p)) = \pm 2(q-p)$. Unfortunately, the expression $\pm 2(q-p)$ can happen to be a multiple of 10. This is the case whenever the difference between the two adjacent digits that are switched is equal to ± 5, for example, if $p = 7$ and $q = 2$, or if $p = 0$ and $q = 5$.

Thus, in contrast to the old ISBN-10 system, the ISBN-13 check digit method cannot catch all transposition errors of adjacent digits. On the other hand, times have changed and typing errors are not such a big problem anymore. Librarians use barcode scanners to read ISBN codes into the computer, and data transfer between publishers to bookstores is not dependent on paper at all, it is fully electronic.

In the late 1960s, when the ISBN system was developed, computers were just starting to be used for business purposes, and only for certain steps of the whole workflow. Numbers were still put into the computer by humans, and humans make a lot of mistakes. In fact, the most common typing mistake that humans make is switching two adjacent digits in a number, the aforementioned transposition error. The idea of using a check digit came about to help find such errors.

HOW THE GLOBAL POSITIONING SYSTEM (GPS) WORKS

The Global Positioning System (GPS) has changed our lives. Navigating a car in an unfamiliar part of town or in a foreign country is not a challenge anymore; you just have to turn on your GPS device, enter your desired destination, and then follow the instructions. In fact, more and more people seem to prefer letting an electronic device figure out the route rather than using old-fashioned maps or activating their memory—even if it's just a short trip for which it would be fairly easy to find the way simply by being willing to read road signs. It is not surprising that parallel to the proliferation of smartphones with GPS technology over the last decade, sales of printed maps have been in rapid decline. Undoubtedly, satellite navigation has made our lives easier. However, there may be some adverse effects as well. You may have noticed while using computer navigation that although it is useful to get from one place to another, it makes it harder to remember where you've been or how you got there. Scientific studies have shown that this sort of navigation is a use-it-or-lose-it skill.[7] Thus, over-reliance on GPS technology could lead to a deterioration of our natural navigation abilities. Moreover, the way in which we typically use GPS navigation distinguishes this technology from other technologies that have become an integral part of our lives. When we obediently follow every command dictated by the automated voice in our navigation system, we reinforce the increasing dependence of our society on computers and artificial intelligence in its most direct form, and perhaps also in its most embarrassing form. Here it becomes blatantly visible that we are not always in control of technology, but, more often, technology controls us. GPS navigation works so well that some drivers start to have more faith in their GPS navigation systems than they do trust their own eyes. They blindly follow the directions, without any doubt or question, even if those instructions were to direct us into a lake, down a cliff, or straight into a house.[8] No technology is infallible; mistakes will always occur. This should not come as a surprise, yet some drivers still seem to switch off their brains as soon as they switch on their GPS navigators. When drivers are in this state of mind, typos within the GPS navigation can be fatal. A pinch of common sense should suffice to help you realize that there might be something wrong when you have already driven hundreds of miles, directed by your navigation system, but all you wanted was to get from the airport to your nearby hotel. Epic detours and other disasters resulting from too much faith in GPS navigation combined with impressive resilience against common sense are reported in the media almost every week. However, in spite of such amusing mishaps, the Global Positioning System remains an extremely useful technology.

As with most technologies developed in the twentieth century, the GPS involves a lot of mathematics. A comprehensive description of all mathematical methods that are applied to reach the remarkable accuracy of this system would go far beyond the scope of this book, but the basic mathematical concepts underlying satellite navigation are actually very simple. We will try to reveal these essential ideas without going into the details of the algorithm that is used to compute the position from the received data. Although the mathematical methods that are necessary to fully understand the computation algorithm do not require much more than high-school mathematics, the formulas involved would only distract from the fundamental ideas and make everything appear more complicated than it actually is, without providing any further significant insights.

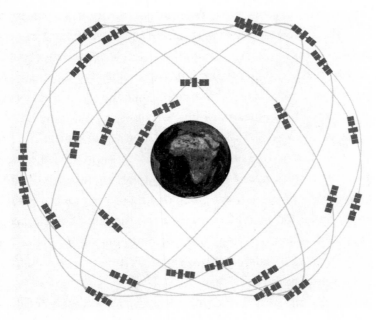

Figure 8.38. (Image from GPS.gov.)

The Global Positioning System (GPS) is operated by the United States Air Force and uses a constellation of (at least) twenty-four satellites, each in its own orbit 11,000 nautical miles[9] (approximately 20,500 kilometers or 12,500 miles) above the earth. The constellation comprises six equally-spaced orbital planes surrounding the earth, where each plane contains the orbits of four satellites (see fig. 8.38). This arrangement ensures that from any point on the planet, at any time, at least four satellites can be viewed. Each satellite permanently broadcasts messages in regular time intervals. These messages are transmitted as microwave signals and contain information enabling a GPS receiver on, or near, the earth's surface to calculate its distance to the satellite and the position of the satellite with respect to the earth.

Each message contains a code identifying the satellite, information about its exact orbit (data called *ephemeris*), and the precise time at which the signal has been emitted. The transmitted data enable the receiver to precisely calculate the current position of the satellite. However, the most important part of the message is the time stamp, measured by clocks onboard the satellite. The distance between the satellite and the receiver can be computed by comparing the time of arrival of the message with its time of emission, since the signal always travels at the same speed, the speed of light, which is approximately 300,000 kilometers per second (or 186,000 miles per second). The time delay between emission and receipt of the signal multiplied by the speed of light gives us the distance of the satellite at the time of emission, which is in very good approximation of the position of the satellite at the time of receipt. Since the satellite moves much slower than the speed of light, its position doesn't change significantly during the signal's traveling time (it takes the signal only a small fraction of a second to travel from the satellite to the receiver).

Incidentally, the same principle can be used to estimate the distance of a nearby thunderstorm by counting the seconds that pass between a flash of lightning and the thunder. The flash of light from the lightning and the sound of thunder are created at the same place and time, but sound travels about one kilometer every three seconds (or one mile every five seconds), which is again much, much slower than the speed of light. That's why we first see the lightning and then hear the thunder afterwards. The sound requires more time to reach us. The farther away we are from the thunderstorm, the longer the delay between light and sound. To calculate your distance from the lightning source, you can use a stopwatch to determine the time delay between lightning and thunder in seconds. Dividing the result by 3, or 5, will give you the approximate distance of the thunderstorm in kilometers or miles, respectively.

Similarly, your GPS device counts the time delay between emission and receipt of a satellite signal and converts it into a distance by multiplying it by the speed of light. At any instant in time, the collection of all points with the same distance d_1 to a satellite S_1 forms a sphere of radius d_1 around this satellite as shown in figure 8.39. The intersection of this sphere with the surface of the earth is a circle C_1 on the surface of the earth. By receiving the signal from satellite S_1 you can, therefore, localize yourself on the circle C_1. If your GPS device also receives a signal from a second satellite S_2 and computes the corresponding distance, d_2, you can localize yourself on the intersection of the circles C_1 and C_2, where C_2 is the intersection of the surface of the earth with a sphere of radius d_2, centered at satellite S_2. As you can see in figure 8.39, two satellites are not sufficient to determine your position uniquely, since two circles on a sphere will generally have two points of intersection. The signal from a third satellite is needed to single out one of the two points and obtain your position as the unique point of intersection of C_1, C_2, and C_3.

Figure 8.39.

Earlier we mentioned that no matter where on the surface of the planet you are, the special arrangement of the orbits of the twenty-four GPS satellites ensures that there are always at least four satellites visible. Why are four satellites needed, when three of them are already sufficient to determine your location? Recall that the GPS receiver computes the traveling time of a satellite signal by comparing its time of emission (which is measured by a clock onboard the satellite and transmitted with the signal) with the time of receipt, which is measured by a clock built into the GPS receiver. If the clock at the receiver were perfectly synchronized with the satellite clock, then the difference between emission in satellite-clock time and receipt in receiver-clock time would indeed be the true traveling time of the signal. However, since the GPS receiver clock is, in general, not perfectly synchronized with the satellite clock, the traveling times computed by the receiver are in error. This is the reason a fourth satellite is needed. All GPS satellites are equipped with very stable atomic clocks that are synchronized with one another and to ground clocks. Any drift from true time maintained on the ground is corrected daily. Because all satellite clocks are perfectly synchronized, the error in the traveling times as they are computed by the receiver can be attributed exclusively to the time offset ($T_0 = T_{Receiver} - T_{Satellites}$) between the receiver clock and the satellite clocks. If c is the speed of light, then the distances computed by the receiver will deviate from the real distances by an amount cT_0 (which can be positive or negative). To locate a point in space, you need three "numbers" or three coordinates. To locate a point on Earth, you have to know its latitude, longitude, and height above sea level. These are called *spherical coordinates*, which are very convenient to use, since the earth is approximately a sphere. An alternative to spherical coordinates are Cartesian coordinates x, y, and z, which are part of the standard high-school curriculum. Since the point we are going to make is independent of which coordinate system we use, we will stick to Cartesian coordinates x, y, and z. Assume that x_k, y_k, and z_k are the spatial coordinates of satellite S_k (at a certain time). By using the Pythagorean theorem, we can write the distance between a point P with coordinates p_x, p_y, and p_z and the satellite S_k as $\sqrt{(p_x - x_k)^2 + (p_y - y_k)^2 + (p_z - z_k)^2}$. If the point P is the position of the GPS receiver and d_k are the computed distances to the satellites, which deviate from the real distances by an offset cT_0, then we get the following system of equations:

$$\sqrt{(p_x - x_1)^2 + (p_y - y_1)^2 + (p_z - z_1)^2} = d_1 + cT_0$$

$$\sqrt{(p_x - x_2)^2 + (p_y - y_2)^2 + (p_z - z_2)^2} = d_2 + cT_0$$

$$\sqrt{(p_x - x_3)^2 + (p_y - y_3)^2 + (p_z - z_3)^2} = d_3 + cT_0$$

There are four unknown quantities in these equations, but since we only have three equations, we do not get a unique solution; that is, we cannot uniquely determine the values of the unknowns—p_x, p_y, p_z, and

T_0—with only the three equations above. To obtain a unique solution of a system of equations with four unknowns, we need four equations. That is why a fourth satellite is needed. Adding the equation

$$\sqrt{(p_x - x_4)^2 + (p_y - y_4)^2 + (p_z - z_4)^2} = d_4 + cT_0$$

to the system of equations above, we have four equations with four unknown quantities. It is exactly this system of equations that your GPS device must solve to determine your location. It is a nonlinear system for four unknowns, which is a bit more complicated to solve than the equations we had in high school, so we shall not distract your attention here by showing how your GPS receiver solves this system of equations.

There is another interesting aspect of the Global Positioning System we ought to mention. As you may know from your own experience with GPS devices, the accuracy of the localization is incredibly high. The calculated position typically differs by no more than 5 to 10 meters in the horizontal. Given that the speed of light is approximately 300,000 km per second, this deviation corresponds to a time span of only about 10 nanoseconds. Thus, to maintain this remarkable precision, the satellite clocks must keep time with an accuracy of 10 nanoseconds per day (deviations from true time are corrected daily from control sites on the ground). An accuracy of 10 nanoseconds per day means that the deviation from true time is only 1 second in 300,000 years. Only atomic clocks can provide this level of accuracy. Since the mechanism by which atomic clocks work is an application of quantum physics, the Global Positioning System would not have been possible without quantum physics. Interestingly, the accuracy of the GPS also relies on the special and general theories of relativity. The satellites are constantly moving relative to observers on the earth, and special relativity predicts that a moving clock is slower than the stationary one, an effect called *time dilation*. General relativity predicts that clocks also go slower in a higher gravitational field. This gravitational effect makes the clock aboard the GPS satellites click faster than a clock down on Earth. Both effects have to be considered and, in total, the clocks on a GPS satellite are faster than a clock on Earth by about 38 microseconds every day, which would be 1,000 times more than the acceptable deviation. This is not a matter of building still more accurate clocks, but realizing that time itself goes faster in the satellites. However, since these effects can be calculated very accurately for a given orbit, they can be compensated by a correction factor. If these relativistic effects were not taken into account, the GPS would accumulate a localization error of about 11 kilometers per day, rather than its current error of only 5 to 10 meters. It is very interesting to note that the two most revolutionary physical theories, and perhaps also the two greatest intellectual achievements of the twentieth century, are both essential ingredients of the Global Positioning System. It would not exist—or at least not work—without them. Unfortunately, the involvement of these masterpieces of science does not prevent people from driving their cars into lakes or down a cliff, just because their GPS told them to drive in that direction!

Having now journeyed with descriptive numbers in our environment, and numbers that helped us identify our location and action in our environment, we comfortably conclude that the reader should now have a rather-comprehensive feeling as to how mathematics accompanies us and supports us through our daily lives.

ACKNOWLEDGMENTS

A book of this complexity requires very capable editorial management and copyediting. We wish to thank Catherine Roberts-Abel for very capably managing the production of this book, and offer very special thanks to Jade Zora Scibilia for truly outstanding editing throughout the various phases of production. We also thank Hanna Etu for her support, which was also much appreciated. The editor in chief of Prometheus Books, Steven L Mitchell, deserves praise for enabling us to approach the general readership to expose the mathematical wonders that help explain our everyday experiences. Christian Spreitzer would like to thank Dr. Katharina Brazda for contributing ideas and for her support during the writing of this book.

NOTES

CHAPTER 1: HISTORICAL HIGH POINTS IN THE DEVELOPMENT OF MATHEMATICAL APPLICATIONS

1. Heinrich Weber, "Leopold Kronecker," *Mathematische Zahlen* 43, no. 1, Berlin/Heidelberg: Springer Verlag, pp. 1–25.

2. Karl Menninger, *Number Words and Number Symbols: A Cultural History of Numbers* (New York: Dover Publications, 1992), p. 399. Originally published Cambridge, MA: MIT Press, 1969.

3. Robert Kaplan, *The Nothing That Is: A Natural History of Zero* (London: Allen Lane/Penguin Press 1999), pp. 68–75.

4. Syamal K. Sen and Ravi P. Agarwal, *Zero: A Landmark Discovery, the Dreadful Void, and the Ultimate Mind* (Amsterdam: Academic Press, 2016), p. 6.

5. Fibonacci used the term "Indian figures" for the Hindu numerals. Laurence Sigler, *Fibonacci's Liber Abaci: A translation into Modern English of Leonardo Pisano's Book of Calculation* (New York: Springer Verlag, 2002).

6. Lucas N. H. Bunt, Phillip S. Jones, and Jack D. Bedient, *The Historical Roots of Elementary Mathematics* (New York: Dover, 1976).

7. Ibid., chap. II.7.

8. Paul Erdős,(1950), "Az $1/x_1 + 1/x_2 + \ldots + 1/x_n = a/b$ egyenlet egész számú megoldásairól" ("On a Diophantine Equation") *Mat. Lapok.* 1:192–210 (PDF, in Hungarian).

9. See, for example, *Encyclopaedia Britannica*, s.v. "Pendulum," https://www.britannica.com/technology/pendulum.

10. H. Sully and J. Le Roy, *Regle artificielle des tems* (Paris: G. Dupuis, 1737), chap. 1, p. 14.

11. See, for example, *Encyclopaedia Britannica*, s.v. "Watch," https://www.britannica.com/technology/watch#ref38408.

12. See, for example, J. J. O'Connor and E. F. Robertson, "Decimal Time and Angles," October 2005, http://www-history.mcs.st-and.ac.uk/HistTopics/Decimal_time.html.

13. Ibid.

14. Darren Rovell, "NFL: It's Super Bowl 50, Not L," ESPN, June 5, 2014, http://www.espn.com/nfl/story/_/id/11031941/nfl-take-one-year-hiatus-roman-numerals-super-bowl-50.

15. Angus Maddison, *Contours of the World Economy, 1–2030 AD* (New York: Oxford University Press, 2007), pp. 32–33.

16. David Eugene Smith, "John Wallis as a Cryptographer," *Bulletin of the American Mathematical Society* 24, no. 2 (1917): 82–96.

17. See, for example, *Wikipedia*, s.v. "Calendar (New Style) Act 1750," https://en.wikipedia.org/wiki/Calendar_(New_Style)_Act_1750.

CHAPTER 2: MATHEMATICS IN OUR EVERYDAY LIVES— ARITHMETIC SHORTCUTS AND THINKING MATHEMATICALLY

1. Two numbers are relatively prime if they have no common factors other than 1.
2. U.S. Code, Title 15, Chapter 6, Subchapter I, § 204.

CHAPTER 4: PROBABILITY, GAMES, AND GAMBLING

1. Victor J. Katz, *A History of Mathematics: An Introduction*, 3rd ed. (Boston: Addison-Wesley/Pearson, 2009), p.490.

2. See, for example, "Five Facts about Friday the 13th," NBC News, March 12, 2009, http://www.nbcnews.com/id/29661652/ns/technology_and_science-science/t/five-facts-about-friday-th/#.Wuy-SIgvwwE; Randy Fishell, *Ready 2Go* (Hagerstown, MD: Autumn House Publishing, 2008), p.11.

3. Bancroft H. Brown, *American Mathematical Monthly* 40 (1933): 607.

4. Ernest Newman, *The Life of Richard Wagner*, 4 vols. (New York: Alfred A. Knopf, 1933); Barry Millington, ed., *The Wagner Compendium: A Guide to Wagner's Life and Music* (New York: Schirmer and Sons, 1992).

5. *Wikipedia*, s.v. "Tic-tac-toe," https://en.wikipedia.org/wiki/Tic-tac-toe.

6. Jason Rosenhouse, *The Monty Hall Problem: The Remarkable Story of Math's Most Contentious Brain Teaser* (Oxford: Oxford University Press, 2009).

7. Ralph Nelson Elliott, *The Wave Principle* (New York, 1938).

8. Ibid.

9. Ralph Nelson Elliott, *Nature's Law: The Secret of the Universe* (New York, 1946).

10. For more information, see Alfred S. Posamentier and Ingmar Lehmann, *The Glorious Golden Ratio* (Amherst: Prometheus Books, 2012).

11. Robert Fischer, *Fibonacci Applications and Strategies for Traders* (New York: John Wiley, 1993).

12. For the arithmetic mean a and b is defined for all real numbers, for the geometric mean it is defined for $a \geq 0$, and $b \geq 0$, and for the harmonic mean a and b are defined for all real numbers $a, b > 0$ and $a + b = 0$.

CHAPTER 5: SPORTS AND GAMES— EXPLAINED MATHEMATICALLY

1. Jason König, "A Comparison: The Modern and Ancient Olympics," *Newsweek*, August 3, 2016, http://www.newsweek.com/comparison-modern-and-ancient-olympics-486793

2. Billie Jean King, http://www.azquotes.com/quote/591896.

3. "The Spirograph: Educational Toy of the Year, 1965–1967," http://workflow.arts.ac.uk/artefact/file/download.php?file=36266&view=5934; Toy Retailers Association, "Toy of the Year, 1965–2007," https://www.toyretailersassociation.co.uk/toy-year-1965-2007.

CHAPTER 6: THE WORLD AND ITS NATURE

1. Carl B. Boyer, *A History of Mathematics* (New York: John Wiley, 1968), pp. 176–79.

2. Kenneth Appel and Wolfgang Haken, "Solution of the Four Color Map Problem," *Scientific American* 237, no. 4 (October 1977): 108–21, doi:10.1038/scientificamerican1077-108.

3. Victor J. Katz, *A History of Mathematics*, 3rd ed. (New York: Addison-Wesley, 2009), pp. 701–702.

4. A. S. Posamentier and I. Lehmann, *The Fabulous Fibonacci Numbers* (Amherst, NY: Prometheus Books, 2007).

5. T. Antony Davis and T. K. Bose "Fibonacci System in Aroids," *Fibonacci Quarterly* 9, no. 3 (1971): 253–55.

6. Roger V. Jean, *Phyllotaxis: A Systemic Study in Plant Morphogenesis* (Cambridge, UK: Cambridge University Press, 1994).

7. These are species of Compositae—one of the largest families of the vascular plants. See P. P. Majumder and A. Chakravati, "Variation in the Number of Rays- and Disc-Florets in Four Species of Compositae," *Fibonacci Quarterly* 14, no. 2 (1976): 97–100.

8. T. Antony Davis, "Why Fibonacci Sequence for Palm Leaf Spirals?" *Fibonacci Quarterly* 9, no. 3 (1971): 237–44.

9. "Des proportions du corps humain," *Bulletin de l'Académie Royale des Sciences, des Lettres et des Beaux-Arts de Belgique* 15 (1848), no. 1, pp. 580–93, and no. 2, pp. 16–27.

10. Adolf Zeising, *Neue Lehre von den Proportionen des menschlichen Körpers* (Leipzig: R. Weigel, 1854).

11. Le Corbusier, *The Modulor: A Harmonious Measure to the Human Scale, Universally Applicable to Architecture and Mechanics* (Basel and Boston: Birkhäuser, 2004). First published in two volumes in 1954 and 1958.

12. Cecie Starr, Christine A. Evers, Lisa Starr, *Biology: Concepts and Applications, Brooks/Cole Biology Series* (Belmont, CA: Thomson, Brooks/Cole, 2006).

13. John C. Manning, *Applied Principles of Hydrology*, 3rd ed. (Upper Saddle Hall, NJ: Prentice Hall, 1996), p. 41.

14. Paul A. Tipler and Gene Mosca, *Physics for Scientists and Engineers*, 6th ed. (New York: Freeman, 2008).

15. Ivor Grattan-Guinness, *The Norton History of Mathematical Sciences* (New York: W. W. Norton, 1998), pp. 230–31; R. A. R. Tricker, *Introduction to Meteorological Optics* (London: Mills and Boon, 1970), chaps. 3–6; M. Minnaert, *The Nature of Light and Colour in the Open Air* (New York: Dover, 1954), pp. 167–90; M. Abramowitz and I. Stegun, *Handbook of Mathematical Functions* (Washington, DC: USGPO, 1964), pp. 446–52, table 10.11, pp. 475–78; H. M. Nussenzweig, *The Theory of the Rainbow*, *Scientific American*, April 1977; J. B. Calvert, "The Rainbow," July 31, 2003, last updated August 14, 2003, https://mysite.du.edu/~jcalvert/astro/bow.htm.

16. Tipler and Mosca, *Physics for Scientists and Engineers*; Color Vision and Art, "Newton and the Color Spectrum," http://www.webexhibits.org/colorart/bh.html.

17. Eugene Hecht, *Optics*, 4th ed. (Reading, MA: Addison-Wesley, 2002).

CHAPTER 7: APPEARANCES OF MATHEMATICS IN ART AND ARCHITECTURE

1. The equality of two ratios is called a *proportion*.
2. We cross multiply, that is, the product of the means equals the product of the extremes.
3. Euclid, *Elements*, book 2, proposition 11; book 6, definition 3, proposition 30; book 13, proposition 1–6.
4. The quadratic formula for solving for x in the general quadratic equation $ax^2 + bx + c = 0$ is $x = \frac{-b \pm \sqrt{b^2 - 4ac}}{2a}$.
5. There is reason to believe that the letter ϕ was used because it is the first letter of the name of the celebrated Greek sculptor, Phidias (ca. 490–430 BCE) [in Greek: (Pheidias) ΦΕΙΔΙΑΣ or Φειδίας], who produced the famous statue of Zeus in the Temple of Olympia and supervised the construction of the Parthenon in Athens, Greece. His frequent use of the golden ratio in this glorious building (see chapter 2.) is likely the reason for this attribution. It must be said that there is no direct evidence that Phidias consciously used this ratio.

    The American mathematician Mark Barr was the first using the letter ϕ in about 1909—see Theodore Andrea Cook, *The Curves of Life* (1914; New York: Dover, 1979), p. 420. It should be noted that sometimes you will also find the lower case φ— or less commonly the Greek letter τ (tau), the initial letter of τομή (to-me or to-mī') meaning *to cut*.
6. A *unit square* is a square with side length of 1 unit.
7. Adolf Zeising (1810–1876), a German philosopher, *Neue Lehre von den Proportionen des menschlichen Körpers* (New Theories about the Proportions of the Human Body) (Leipzig, Germany: R. Weigel, 1854). The book *Der goldene Schnitt* (The Golden Ratio) was published posthumously (Halle, Germany: Leopoldinisch-Carolinische Akademie, 1884).
8. Gustav Theodor Fechner, *Zur experimentalen Ästhetik* (On Experimental Aesthetics) (Leipzig, Germany: Breitkopf & Haertel, 1876).
9. Luca Pacioli, *De divina proportione* (Venezia, 1509; reprinted in Milan in 1896; Gardner Pelican, 1961).
10. Cf. the discussions at the beginning of the chapter 7.2, Sculptures.
11. Dan Pedoe, *Geometry and the Visual Arts* (New York: Dover, 1976).
12. This actually refers to Giotto di Bondone, Italian painter and master builder.

13. Masaccio is Tommaso di Ser Giovanni di Simone; he is the Italian painter who is regarded as a founder of Renaissance painting. The Trinity fresco was of pioneering meaning for the mural painting and the development of the altar because of the central perspective.

14. He was also called Roger de Le Pasture, a Flemish painter.

15. Pietro di Benedetto dei Franceschi, also Pietro Borgliese, Italian painter.

16. Pietro Perugino was born Pietro Vannucci.

17. Michelangelo Buonarroti; Michelangelo was, however, a sculptor, painter, architect, and poet. In 1508, Michelangelo also designed the uniforms of the Swiss guard for the Vatican.

18. Italian copperplate engraver.

19. *Painting* I and *Composition with Colored Areas and Gray Lines* 1 (1918) e.g. are often quoted; as well *Composition with Gray and Light Brown* (1918; Museum of Fine Arts, Houston, Texas), *Composition with Red Yellow Blue* (1928).

20. Actually José Victoriano González-Pérez, Spanish painter and graphic artist.

21. German painter, graphic artist, and wood engraver.

22. J. J. O'Connor and E. F. Robertson, "Joachim Jungius," May 2010, http://www-groups.dcs.st-and.ac.uk/history/Biographies/Jungius.html.

23. *Wikipedia*, s.v. "Catenary," https://en.wikipedia.org/wiki/Catenary.

24. Thomas Paine to Thomas Jefferson, September 9–15, 1788, https://founders.archives.gov/documents/Jefferson/01-13-02-0466.

25. *Wikipedia*, s.v. "Gary Anderson (designer)," https://en.wikipedia.org/wiki/Gary_Anderson_(designer).

CHAPTER 8: THE TECHNOLOGY AROUND US— FROM A MATHEMATICAL PERSPECTIVE

1. Clifford A. Pickover, *The Math Book* (New York: Barnes & Noble, 2009), p. 504; *Wikipedia*, s.v. "Britney Gallivan," https://en.wikipedia.org/wiki/Britney_Gallivan.

2. G. J. Toomer, *Diocles on Burning Mirrors* (New York: Springer-Verlag, 1976).

3. Henry Petroski, *The Road Taken: The History and Future of America's Infrastructure* (New York: Bloomsbury USA, 2017).

4. *Wikipedia*, s.v. "Caesar cipher," https://en.wikipedia.org/wiki/Caesar_cipher.

5. *Wikipedia*, s.v. "Observable universe," https://en.wikipedia.org/wiki/Observable_universe.

6. *Wikipedia*, s.v. "International Standard Book Number," https://en.wikipedia.org/wiki/International_Standard_Book_Number.

7. E. A. Maguire et al., "Navigation-Related Structural Change in the Hippocampi of Taxi Drivers," *Proceedings of the National Academy of Sciences USA* 97 (2000): 4398–403.

8. Lauren Hansen, "8 Drivers Who Blindly Followed Their GPS into Disaster," *Week*, May 7, 2013, http://theweek.com/articles/464674/8-drivers-who-blindly-followed-gps-into-disaster.

9. 1 nautical mile = 1,852 m or 6,076.1 ft.

INDEX

NUMBERS

0. *See* zero (0)
0.2360. *See* golden ratio (ϕ)
2
    irreducible fraction with 2 as the numerator, 27
    rules for divisibility, 62–63
    shortcut for a unit fraction for irreducible fractions with a numerator of 2, 26–27
    square root of 2 ($\sqrt{2}$)
        and paper sizes, 108–109
        and wrapping packages, 133
        use of in ancient Egyptian multiplication, 23–25
2.71828182845904523536028747135 27 . . . *See* Euler's constant (*e*)
4
    colors needed to color any map, 233–37
    number of satellites always visible for GPS, 383, 386
5
    rules for divisibility, 62–63
    squaring numbers with terminal digit 5, 69–71
7 and rules for divisibility, 63–65

9
    "casting out nines" to check arithmetic, 60–62
    multiplying any number by, 59
    rules for divisibility, 54–56

10
    apostrophus notation system representing multiplication by 10s, 38
    base-10 number system, 15, 30, 31, 62, 76
        and base-60 system, 18, 32
        introducing zero into, 19
        timekeeping not using, 29
    multiplication by factors of powers of 10, 68–69

11
    multiplying two-digit numbers by, 57–58
    as a prime number, 379
    properties of, 54
    rules for divisibility, 56–57, 379–80

13
    as a lucky (or unlucky) number, 135–37
    rules for divisibility, 65–66
17 and rules for divisibility, 66–67

401

42°, rainbows appearing under an angle of, 269, 272

60
- base-60 numbering system
  - and Babylonian mathematics, 18, 29–33
  - and base-10 system, 18, 32
  - handling fractions, 32
  - measurements based on integer factors of 60, 32
  - in timekeeping, 29, 33
  - use of a symbol for zero, 19
- divisibility properties of, 32, 33

72, rule of, 104–106

114, rule of, 106

SYMBOLS

e. *See* Euler's constant (e)
ϕ. *See* golden ratio (ϕ)
∞. *See* infinity
π. *See* pi (π)
$\sqrt{}$. *See* square roots

NAMES AND TERMS

accounting principle, fundamental, 142
acoustic waves, 335, 336, 341, 342, 343, 344
Adam and Eve (Dürer), 295
Adam and Eve (Raimondi), 295
Adams, John, 140
Adams, John Quincy, 336
addition
- adding two fractions, 88
- and "casting out nines," 60–62
- finding reciprocals when you know product and sum of two numbers, 88
- finding the weighted sum, 378–79, 380
- and numbers 9 and 11, 54–57
- in Roman mathematics, 37
- sum of two like parities, 91

Adoration of the Magi, The (Leonardo da Vinci), 299
ages, determining for three sons, 74–75
Alberti, Leon Battista, 296
algebraic comparisons of three means, 175–76
Ambassadors, The (Holbein, the Younger), 304
Amenemhat III (ancient Egyptian king), 22
American Mathematical Monthly (journal), 136
American National Standards Institute (ANSI), 106, 107, 109
anchor escapement, 28–29
angles
- angle-dependent coordinates in an epitrochoid, 215
- angle of clock hands in displays, 281–83
- mathematical strategies for billiards, 201–206
- optimizing the soccer shot, 187–92

rainbows always appearing under an angle of 42°, 269, 272
tennis as game of angles, 192–201
See also deflection angle (ϕ); incidence, angle of; reflections, use of; refraction, angle of
ANSI. *See* American National Standards Institute (ANSI)
antisolar point, 272
apostrophus notation system, 38
Appel, Kenneth, 233–34, 237
Arabic numbers. *See* Hindu-Arabic numbering systems
Archimedes, 349
architecture, mathematical applications in, 12–13, 275
    catenary curves and parabolas as curves, 312–15
    finding the center of gravity for building a tower, 333
    and the golden ratio, 12, 160, 276, 279, 284, 287, 294
    human body as basis for architecture, 255, 284, 365–66
    importance of area in home construction, 114–15, 116
    whispering galleries, 334–44
areas
    basic units of in base-60 number system, 32
    of a circle with radius of 1, 116
    comparing areas and perimeters, 110–13
    comparing circle and Reuleaux triangle, 122–23

$\cos a \cdot \sin a$ as area of a rectangle inscribed in a quarter circle, 186
    and designing a coffee-cup sleeve, 127
    and paper folding, 322, 323
    sizing proper ductwork, 111, 114–15
    and wrapping packages, 128, 131–32
arithmetic
    in ancient Egypt, 21–28
    arithmetic shortcuts, 53–92
    in Babylon, 29–33
    historical highpoints, 15–52
    mental arithmetic, 73–74
    in Roman mathematics, 37–38
    solving problems by considering extremes, 82–86
    and thinking "outside the box," 79–82
arithmetic mean, 105, 170, 172–73, 326
    and centers of gravity, 331–32
    comparing three means algebraically, 175–76
    comparing three means geometrically, 177–79
arithmetic progression, 326
art, mathematical applications in, 12–13, 275–318
    catenary and parabola as curves, 312–15
    golden ratio in, 12, 276–81, 284–96
    golden rectangles in, 281–83, 286, 288, 292–93

404 INDEX

Möbius strip as art, 315–18
numbers in art, 305–309
perspectivity, 296–304
viewing a statue optimally, 309–312
Arthur, Chester A., 228
attendance, determining number of people in, 81
average. *See* mean (average); median (average); mode (average)

Babylonian numbering system, 17, 18–19, 29–33
bad luck and frequency of Friday the 13th, 135–37
ballistic curve, 186
ball throwing, best angle for, 181–87
Baptism of Christ, The (Piero della Francesca), 294
Barbara (Saint), 289–90
base-10 number system. *See* decimal system (base-10 number system)
base-60 numbering system. *See* sexagesimal numbering system (base-60)
baseball batting averages, 170–71
Bathers at Asnières (Seurat), 292–93
bees, Fibonacci numbers seen in, 244–45
Bellaso, Giovan Battista, 369–70
Bernoulli, Johann, 313
bicycles and mathematics, 206–211
billiards, mathematical strategies for, 201–206
birthdates, unexpected birthday matches, 137–40

Birth of Venus, The (Botticelli), 286–87
Blaschke, Wilhelm, 123
books, identifying through ISBN numbers, 376–81
Botticelli, Sandro, 286–87
bottle problems
concentrations of red and white wine, 84–85
measuring 7 liters with only 11- and 5-liter cans, 88–91
boxes, sizing for greatest volume, 115–16
Brahmagupta, 19
Brāhmasphuṭasiddhānta, 19
bridges
crossing of, 237–41
curves found in, 312–13
Brown, Bancroft H., 136
Brunelleschi, Filippo, 296
business applications of mathematics, 163
Fibonacci numbers and the stock market, 158–63
interest rates, 100–106
mathematics of life insurance, 163–67
use of discounts, 95–97

Caesar's code or cipher, 367–68, 370
calendar and mathematical applications, 39–52
change from Julian to Gregorian in 1752, 40
determining day for a date in any year, 43–47

finding date for Easter Sunday, 47, 51–52
frequency of the 13th falling on Friday, 135–37
George Washington's birthday, 52
methods for determining day of any given date or holiday, 40–47, 50
    finding dates with same day of the week, 48–49
peculiarities found in, 48–51
probability of having same birthdate, 137–40
problem of perpetual calendars, 47, 51
reusing an old calendar, 49
Calling of Saint Matthew, The (Caravaggio), 303
Caravaggio, Michelangelo Merisi da, 303
Carbajal, Enrique (Sebastián), 317
cardioid, 354, 355–56
cards, playing
    probability of matching two, 142–44
    probability of various hands in poker games, 144–50
cars, problems involving
    coordination of traffic signals, 356–65
    determining speeds of cars, 85–86, 320, 359, 362, 363, 364
    distance necessary for a car to overtake another with a head start, 320

how Global Positioning System works, 381–89
which gets wetter, a slower or faster car, 82–83
Cartesian coordinates, 387
"casting out nines," 60–62
catenary as a curve, 312–13
caustics, 351–56
center of gravity and building a skewed tower, 328–34
central tendency, measure of, 170, 172
    comparing three means algebraically, 175–76
    comparing three means geometrically, 177–79
centroid, 333
chains of unit fractions, long, 27–28
check-digit method for ISBNs, 377–81
cipher algorithms, 367
    Caesar's code or cipher, 367–68, 370
    Vigenère cipher, 369–70
ciphertext, 367, 368, 370
circles
    and caustics in a coffee cup, 351–56
    coffee-cup sleeve as a segment of a circular ring, 126–28
    comparing areas and perimeters, 112–13
    comparing circle and Reuleaux triangle, 120, 122–23
    concentric circles, 112
    congruent circles, 337
    Euler labeling ratio of diameter

INDEX

and circumference as π (pi), 144. *See also* pi (π)
full-circle rainbows, 269, 272
as perfect manhole covers, 118
and the Spirograph toy, 212–23
topologists' view of, 233
two circles with same diameter, 354, 355–56
walking a mile at the North and South Poles, 229–32
Circus Sideshow (Seurat), 292
clocks and watches
angle of clock hands in displays, 281–83
history of timekeeping, 28–29
importance of atomic clocks for GPS, 386–87, 388–89
and sexagesimal numbering system, 29, 33
use of correction factor to determine movement of hands of a clock, 319–22
use of roman numerals, 34, 36–37
See also timekeeping
clothes and mathematical problems
best deals in buying, 95–100
selecting clothes, 141–42
codes and messages, creating, 365–76
keyword for translating a coded message, 371–76
coffee mugs
concentrations of light in (caustics), 351–56
designing a coffee-cup sleeve, 124–28

collimated light, 345, 346, 350
Collins, Charles, 159
colors
coloring a map, 233–37
as a factor in perspective, 298
picking socks of a certain color, 141–42
in a rainbow, 259, 271–72
Commissioners 1958 Standard Ordinary Mortality Table, 164, 165
composite numbers and divisibility, 32, 67
compound interest, 100–104
and life insurance, 166–67
rule of 72, 104–106
concave mirrors, 345
concentric circles, 112
congruent circles, 337
construction and mathematics
comparing areas and perimeters, 110–13
perfect manhole cover, 117–24
problems found in the home, 113–17
consumer symbol, Möbius strip as, 315–18
Cook, Theodore, 286
correction factor, 321–22, 389
counting using tally marks, 15–17
coupons, evaluating of, 98–100
Crucifixion (with the Virgin Mary, Saint Jerome, Mary Magdalene, and John the Baptist) (Raphael), 295
cryptography, 365–76

cube, 115
    nets of, 130
cubes of numbers in a magic square, 308
cuboids
    in a skewed tower, 328
    wrapping efficiently, 128–34
curves, 13
    ballistic curve, 186
    breadth of the curve, 120
    catenary curve, 312–15
    light reflected by a curved surface (caustics), 351–56
    parabolas, 313
        and best angle for throwing a ball, 182–83, 186–87
        definition of, 341–42
        in flashlights, 345–48
        in satellite dishes, 350–51
    produced by Spirograph, 213–14, 217–21
        number of orbits to achieve complementary curve, 222
    roulette family of curves, 213
    See also cardioid; epitrochoids; hypotrochoids; nephroids; rainbows, geometry of; roulettes (family of curves); whispering galleries
cycling. *See* bicycles and mathematics
cycloids, 221. *See also* epicycloids; hypocycloids

dates. *See* calendar and mathematical applications
David (Michelangelo), 311
deaths
    death dates of presidents, 140
    and life insurance, 163–67
decimal system (base-10 number system), 15, 30, 31, 62, 76
    and base-60 system, 18, 32
    timekeeping not using, 29
    use of zero, 19–20
Declaration of Independence, 140
decomposition
    in ancient Egyptian mathematics, 23–24, 25, 27
    of nets, 131
    prime number cannot be decomposed, 379
decryption, 367, 368, 370, 371, 375
Dedekind, Richard, 17
De divina proportione (*On the Divine Proportion*) (Pacioli), 284
    illustrated by Leonardo da Vinci, 289
deflection angle (ϕ), 263, 264, 265, 266
denominator, 21
    in ancient Egyptian mathematics, 24, 26, 27
Deposition from the Cross (Weyden), 294
depth perception. *See* perspectivity
Derailleur bicycle drivetrain, 207
De Ratiociniis in Ludo Aleae (Huygens), 135
Der Freischütz (Weber), 137
Der Goldene Schnitt (Zeising), 279

Descartes, René, 266, 273
"Descartes ray," 266
dice game, 135
Diffie-Hellman-Merkle method for translating a coded message, 372–76
Diocles, 350
Dionysius' Procession relief (Villa Albani, Rome), 294
discounts
    and effect of successive percentages, 94–98
    and sales-promotion coupons, 98–100
discrete logarithms, 375, 376
distance
    installing an outlet with shortest distance for connecting two lamps, 116–17
    origins of measures of, 76–77
    using Fibonacci numbers to convert kilometers to and from miles, 75–79
    See also speed
divergence, 260
divergent harmonic series, 333
division, 22–23
    by a large prime number, 66–67
    in modulo operations, 373
    rules for divisibility
        for 2, 62
        for 5, 62–63
        for 7, 63–65
        for 9, 54–56
        for 11, 56–57, 379–80
        for 13, 65–66
        for 17, 66–67
    divisibility of composite numbers, 67
dodecahedron, nets of, 130
Doni Tondo, The (Michelangelo Buonarroti), 295
double rainbows. *See* secondary rainbows
Dow theory, complement to, 159
ductwork, sizing of, 111, 114–15
Dürer, Albrecht, 12
    books on measurements, 130, 285, 301
    Dürer conjecture, 130
    and magic squares, 305–309
    and perspectivity, 301–302
    use of golden ratio, 285–86, 295

Earth, 225–43
    coloring a map of, 233–37
    measuring, 225–29
        accuracy of early measurements, 226–27
        determining distance to the horizon, 227–29
    navigating the globe, 229–32
        how Global Positioning System works, 381–89
    See also nature and mathematics
Easter Sunday, finding date for, 47, 51–52
"Echoes of Statuary Hall, The" (*New York Times*), 336
Educational Toy of the Year Award, 212

INDEX 409

Egypt, ancient
   arithmetic in, 21–28
      division, 25–26
      "greedy algorithm for Egyptian fractions," 26–28
      handling fractions, 24–28
      multiplication, 23–25
      numbering system, 17
   use of 12-hour cycles, 33
Einstein, Albert, 232
electric outlets, locating, 116–17
Elliott, Ralph Nelson, 158–59, 160, 162, 163
encryption. *See* secret messages, creating
ephemeris, 384
epicycloids, 221, 354–55, 356
epitrochoids, 213, 214, 215, 216, 221, 222
   examples of, 217, 218, 219, 223
Eratosthenes, 226–27
Erdös, Paul, 27–28
escapement, 28–29
Euler, Leonhard, 143–44
   and the Königsberg Bridges Problem, 239
Euler's constant (e), 143–44
exponentiation to generate keyword for code translation, 372
extremes, use of to solve problems, 14, 83–84
   concentrations of red and white wine, 84–85
   determining speeds of cars, 85–86
   evaluating coupons, 98–99
   matching socks, 141–42

   and Monty Hall problem, 156–58
   sizing proper ductwork, 114
   which car gets wet faster, 82–83
Eyn schön nützlich Büchlein und Underweisung der Kunst des Messens ("A Nice Useful Booklet and Instruction of the Art of Measurement") (Rodler), 301

Fabulous Fibonacci Numbers, The (Posamentier and Lehmann), 244
factors
   measurements based on integer factors of 60, 32
   multiplying factors of powers of 10, 68–69
   of a number or of a multiplication, 21
   prime factors, 63, 67, 379
Fechner, Gustav, 279–81
Feininger, Lyonel, 295
Fermat, Pierre de, 135
Fibonacci (Leonardo of Pisa), 19, 20–21, 60
   author of *Liber Abaci*, 19, 20, 21, 26, 60, 244
Fibonacci Applications and Strategies for Traders (Fischer), 162–63
Fibonacci numbers, 13, 20–21
   and Egyptian fractions, 26–27
   Fibonacci percentages, 161–62
   and the golden ratio, 13, 77, 160, 255, 257, 281, 292
      approximations, 77, 162, 256, 296

found in the human body, 255–57
Le Corbusier's use of, 256–57
in nature
male bee's family tree, 244–45
in plants, 13, 246–55
and the stock market, 158–63
use of zero, 19, 20
using to convert kilometers to and from miles, 75–79
Fillmore, Millard, 140
fire hydrants and the Reuleaux triangle, 121–22
Fischer, Robert, 162–63
Fisher, Denys, 212
"Five-Bedroom-House Problem," 241–43
five different cards, probability of in poker games, 149, 150
flashlight, 345–48
flowers and Fibonacci numbers
leaf arrangements, 250–55
spirals in Aroids, 249–50
flush, probability of in poker games, 146, 150
Ford Model T car, 357
Four Books on Measurement (Dürer), 130, 285, 301
four of a kind, probability of in poker games, 145–46, 150
fractions
adding two fractions, 88
in ancient Egypt, 24–28

in Babylon, 19
comparing two fractions, 94
and determining unit costs, 93–94
irreducible fraction with 2 as the numerator, 27
and the sexagesimal numbering system, 32
François I (king of France), 288
Friday the 13th, 135–37
Friðfinnsson, Hreinn, 293–94
full house, probability of in poker games, 146, 150
fundamental accounting principle, 142
furlong, 76

Galileo Galilei, 285, 313, 315
Gallican, Britney C., 324–25
gambling. *See* probability
games and toys
dice game, 135
Nim, 87–88
playing cards
probability of matching two, 142–44
probability of various hands in poker games, 144–50
Spirograph toy, 212–23
tic-tac-toe, 150–54
See also sports, mathematical applications in
"gardener's ellipse," 339
Gateway Arch (St. Louis, MO), 314
Gauss, Carl Friedrich, 51
gears and gear ratios in bicycles, 206–211
Gelmeroda (Feininger), 295

GeoGebra (software), 223
Geometer's Sketchpad (software), 223
geometrical optics and rainbows, 258–73
geometric mean, 170, 172, 174–75
    comparing three means algebraically, 175–76
    comparing three means geometrically, 177–79
geometric progression, 257, 323, 327
Geometry and the Visual Arts (Pedoe), 287
George Washington Bridge, 312
Giotto, 294
Girl with the Ermine, The (Leonardo da Vinci), 295
Global Positioning System (GPS), 381–89
globe and mathematical applications. *See* Earth
Golden Gate Bridge, 312
golden ratio (ϕ), 275, 276–81
    in architecture, 12, 160, 275, 276, 284, 287, 294
    in art, 12, 284–96
    examples of, 12
    and Fibonacci numbers, 13, 77, 160, 255, 257, 281, 292
        approximations, 77, 162, 256, 296
        found in the human body, 255–57
        Le Corbusier's use of, 256–57
    using to predict stock prices, 160–61, 162
golden rectangle, 12, 162, 278, 279–81
    and angle of clock hands in displays, 281–83
    in art, 286, 288, 292–93
golden spiral, 293
golden triangles, 288–89
GPS (Global Positioning System), 381–89
Grand Central Station (New York City), whispering gallery effect, 342
Grant, Ulysses S., 228
graph theory and the Königsberg Bridges Problem, 237–41
Great Crash of 1929, 158
"greedy algorithm for Egyptian fractions," 26
"green wave" (coordinated traffic lights), 356–65
Gregorian calendar, 39–40, 43, 136
Gregory XIII (pope), 40
Gris, Juan, 296
Guinness Book of World Records Hall of Fame, 151

Haines Fall, NY, 227–28
Haken, Wolfgang, 234, 237
Harding, Warren G., 138
harmonic mean, 168–69, 170, 172, 173–74
    comparing three means algebraically, 175–76
    comparing three means geometrically, 177–79

harmonic series, 333
Hindu-Arabic numbering systems, 21, 34, 60
Holbein, Hans (the Younger), 304
Holy Family with St. John the Baptist, The (Michelangelo Buonarroti), 295
home-construction problems, 113–17
Hooke, Robert, 28–29
Hoover, Herbert, 136
horizon, determining distance to, 227–29
hoses filling a tub, time needed, 116
Huygens, Christiaan, 28–29, 135, 313
hypocycloids, 221
hypotrochoids, 213, 214, 217, 219, 221, 222, 223
    examples of, 214, 220, 221, 223

Iamblichus of Chalcis, 172
icosahedron, nets of, 130
impact parameters, 265
incidence, angle of
    law of reflection (angle of reflection equal to angle of incidence), 262, 263, 346, 353
    in raindrops, 264–66
    Snell's law (relationship of angle of refraction and angle of incidence), 261, 262
incident rays, 262, 263, 265, 267–68, 269, 347, 352
India, 15, 60
    and development of zero, 19, 20
    numbers system, 77
    See also Hindu-Arabic numbering systems

"Infinite Corridor" at MIT, 325
infinity (∞), 38, 98, 103, 333
instantaneous compounding of interest, 103–104
Instructions for Measurement with Compass and Ruler (*Underweysung der Messung mit dem Zirckel und Richtscheyt*) (Dürer), 130, 301
insulation and design of a coffee-cup sleeve, 124–28
insurance, life. *See* life insurance, mathematics of
integer factors of 60, 32
interest rates, 100–104
    rule of 72, 104–106
International Organization for Standardization (ISO), 106–107, 110, 376
International Standard Book Number. *See* ISBN system
irrational numbers, 144
ISBN system, 376–81
ISO. *See* International Organization for Standardization (ISO)
isosceles triangles with base and sides are in golden ratio. *See* golden triangles

Japanese numbering system, 16, 17
javelin throwing, 187
Jean, Roger V., 250
Jeanneret-Gris, Charles-Édouard. *See* Le Corbusier
Jefferson, Thomas, 140, 313
Julian calendar, 39–40, 45

Julius Caesar
    and the Julian calendar, 40
    use of secret codes, 367–68, 370
Jungius, Joachim, 313

Kepler, Johannes, 285
keyword for translating a coded message, 371–76
    Diffie-Hellman-Merkle method, 372–76
kilometers (km)
    converting to and from miles, 13, 75–79
    diameter of sun and distance from Earth in km, 260
    formula for determining distance of a thunderstorm in km, 384–85
    GPS satellites orbit distance from Earth in km, 383
    speed of light or sound in km, 384
King, Billie Jean, 193
Kircher, Athanasius, 336–38
Königsberg Bridges Problem, 237–41
Kronecker, Leopold, 17

large numbers
    how number systems handle, 18, 19, 38, 372, 376
    law of large numbers, 164
lasers, 345
Last Supper, The (Leonardo da Vinci), 12, 295, 297–98
leaf arrangements and Fibonacci numbers, 250–54

leap years, 40, 41–42, 43, 44–46, 49, 136, 138
Le Corbusier, 256–57
ledger/tabloid paper. *See* paper sizes
legal paper. *See* paper sizes
Lehmann, I., 244
Leibniz, Gottfried, 313
length
    basic units of in base-60 number system, 32
    and building a skewed tower, 329, 332, 333
    and golden ratios, 162, 276, 279
    metric measures of, 76
    and paper folding, 324, 325, 326, 327
    and paper sizes, 106, 108
    stadia as a measure of, 227
    tangent length, 191
    of a tennis court, 194, 196
    and wrapping packages, 132, 133
Leonardo da Vinci, 12
    examples of use of golden ratio, 295
    Mona Lisa, 287–89
    illustrating *De divina proportione* (Pacioli), 289
    and perspectivity, 297–98, 299
        distorted images, 303–304
    Vitruvian Man, 255, 284
    writing in reverse direction, 365–66
Leonardo of Pisa. *See* Fibonacci (Leonardo of Pisa)
Let's Make a Deal (TV show), strategy for winning, 14, 150–54
letter paper. *See* paper sizes

Liber Abaci (Fibonacci), 19, 20, 21, 26, 60, 244
Lichtenberg, Georg Christoph, 108
life insurance, mathematics of, 163–67
light
    caustics (light concentrations), 351–56
    collimated light and flashlights, 345–48
    and rainbows, 258–73
    speed of light, 261, 384, 385, 387, 388
    See also light and shadow in art
light and shadow in art, 302–303
lightning/thunder and location of thunderstorms, 384–85
liquids, problems involving
    concentrations of red and white wine, 84–85
    measuring 7 liters with only 11- and 5-liter cans, 88–91
Listing, Johan Benedict, 315
Liszt, Franz, 137
logarithms
    discrete logarithms, 375, 376
    logarithmic spiral, 163
    logarithm to the base, 372, 375
    natural logarithms, 103
logic
    and arithmetic shortcuts, 74–75
    and reverse reasoning, 88
    selecting clothing, 141–42
    of tic-tac-toe, 150–54
    See also problem-solving strategies

Madonna Alba (Raphael), 290–91
Madonna and Child (Perugino), 295
magic squares, 12, 305–309
Man Drawing a Lute (Dürer), 301
manhole cover, perfect, 117–24
map, coloring a, 233–37
Masaccio, 294
mass
    basic units of in base-60 number system, 32
    and building a skewed tower, 329
mathematics, teaching of, 11–12
mean (average), 168–70
    and calendar peculiarities, 48
    and measures of central tendency, 170, 172
    comparing three means algebraically, 175–76
    comparing three means geometrically, 177–79
    Pythagorean means, 172
    See also arithmetic mean; geometric mean; harmonic mean
measurements
    measures of distance, 76–77, 227 (*see also* kilometers [km]; length; miles)
    ratios of basic units of length, area, and mass in base-60, 32
    using Fibonacci numbers to convert kilometers to and from miles, 75–79
    See also central tendency, measure of
median (average), 170

Melencolia I (Dürer), 12, 305–309
mental arithmetic, 73–74
metric system, 76–77. *See also* kilometers
Michelangelo Buonarroti, 295, 311
miles
    converting to and from kilometers, 13, 75–79
    formula for determining distance of a thunderstorm in miles, 384–85
    speed of light or sound in miles, 384
minutes and timekeeping, 28, 29
MIT, 325
Möbius, August Ferdinand, 315
Möbius strip, 315–18
mode (average), 170
modulo operation, 373–75
Modulor, The: A Harmonious Measure to the Human Scale Universally Applicable to Architecture and Mechanics (Le Corbusier), 256–57
Mondrian, Piet, 295, 296
Monroe, James, 140
Monty Hall problem, 14, 154–58
Monty Hall Problem, The: The Remarkable Story of Math's Most Contentious Brain Teaser (Rosenhouse), 158
multiplication, 22–23
    by 9, 59
    by 11, 57–58
    in ancient Egypt, 22–25
    apostrophus notation system, 38
    checking multiplication by "casting out 9s," 60–62
    cross-multiplying two fractions to compare them, 94
    factors of a multiplication, 21
    by factors of powers of 10, 68–69
    finding reciprocals when you know product and sum of two numbers, 88
    finding the weighted sum, 378–79
    squaring numbers with terminal digit 5, 69–71
    of two-digit numbers, 73–74
    of two-digit numbers less than 20, 71–73
Museum of Modern Art (New York City), 125–26

Napoleon Bonaparte, 136
National Football League, 34
National Statuary Hall in US Capitol, 335–36, 339–40
natural numbers, 17
    as congruent modulo 7, 374
    and Fibonacci numbers, 78, 79
    and the golden ratio, 256–57
nature and mathematics, 243–73
    Fibonacci numbers, 243–57
        and the human body, 255–57
        male bee's family tree, 244–45
        in plants, 13, 246–55
    geometry of rainbows, 258–73
    See also Earth
Nature's Law: The Secret of the Universe (Elliott), 159
navigation, 67–65
    how Global Positioning System works, 381–89
    navigating the globe, 229–32

nephroids, 355
"net" of the polyhedron, 128
    of five Platonic solids (convex polyhedrons with congruent faces), 130
    of a rectangular cuboid, 130–32
networks
    and the "Five-Bedroom-House Problem," 241–43
    and the Königsberg Bridges Problem, 237–41
Neveux, Marguerite, 296
New York City traffic lights, 357–58, 364
New York Times (newspaper), 151, 336
"Nice Useful Booklet and Instruction of the Art of Measurement, A," *Eyn schön nützlich Büchlein und Underweisung der Kunst des Messens* (Rodler), 301
Nim (game of), 87–88
Nislen, Tobias, 338
North Pole, 225, 229, 230, 232
"Noughts and Crosses." *See* tic-tac-toe, mathematical logic of
Numa Pompilius, 39
numbers and numbering systems
    comparison of different systems, 17
    factors of a number, 21
    how number systems handle large numbers, 18, 19, 38, 372, 376
    origin of number symbols, 15–17
    tally marks, 16
    place-value systems of numbering, 18, 19, 30, 31–32
        not used by Romans, 30, 36, 38–39
    zero, 17–20
    See also Babylonian numbering system; Egypt, ancient, numbering system; Hindu-Arabic numbering systems; roman numerals
numerator, 21
    in ancient Egyptian mathematics, 24, 26, 27

Olympic games, use of roman numerals, 34–35
On Burning Mirrors (Diocles), 350
one pair, probability of in poker games, 148–49, 150
ordering, retention of, 233
outlet installation, 116–17

Pacioli, Luca, 284, 289
Paine, Thomas, 313
pairs, probability of in poker games, 148–49, 150
Pankok, Otto, 296
paper folding, mathematics of, 322–28
paper sizes, 106–110
    comparing areas and perimeters, 110–13
Pappus of Alexandria, 177
parabolas and parabolic reflectors
    and best angle for throwing a ball, 182–83, 186–87
    definition of a parabola, 341–42

INDEX 417

differentiating parabola and catenary curves, 313, 315
in flashlights, 345–48
parabolic reflectors for amplification of sound, 341
in satellite dishes, 350–51
parabolic concentrators, 348–49
Parade (magazine), 150, 151
parallel light rays
   and caustics in a coffee cup, 352, 353, 356
   collimating property of parabolas, 345, 346, 350
   and a flashlight, 345, 346–48
   in rainbows and raindrops, 260, 261, 264, 267, 269
   use of to determine circumference of the earth, 226–27
parallel lines tangent to a curve. *See* curves, breadth of the curve
parallel radio waves, 350
Parigi, Giulio, 349–50
parity, 91
Parthenon, 12
Pascal, Blaise, 135
Pedoe, Dan, 287
percentages
   combining coupons, 99–100
   effect of successive percentages, 94–98
   fixed percentage coupons, 98–99
   rule of 72, 104–106
   understanding interest rates, 100–104

perimeters
   comparing areas and perimeters, 110–13
   finding largest area from given perimeter, 115
   in "gardener's ellipse," 339
   in the Reuleaux triangle, 122
perpetual calendars, problem of, 47, 51
perspectivity, 12, 275, 286, 296–304
   depth perception, 12, 275, 296, 297, 300, 304
   perspectival anamorphosis, 303–304
   use of light and shadow, 302–303
Perugino, Pietro, 295
Phonurgia Nova (Kircher), 336–37
phyllotaxis. *See* leaf arrangements and Fibonacci numbers
pi (π), 144, 209
   ancient Egyptians calculating, 22
Piero della Francesca, 294
pineapples and Fibonacci numbers, 246
pine cones and Fibonacci numbers, 247–49
place-value systems of numbering, 18
   and Babylonian mathematics, 30, 31–32
   in India, 19
   not used by Romans, 30, 36, 38–39
plaintext, 367, 368, 370
plants and Fibonacci numbers, 246–55
Plimpton 322 (clay tablet), 29–30
Poincaré, Henri, 33

poker and mathematical applications, 14, 142
   probability of various hands in poker games, 144–50
Polk, James K., 138
polyhedrons, nets of, 128–29, 130–32
pool game. *See* billiards, mathematical strategies for
Poor Richard's Almanack (Franklin), 40–41
Posamentier, A. S., 244
powers of 10, 18
   multiplication of powers of 10, 68–69
presents, optimally wrapping, 128–34
prime factors, 63, 67, 379
Principle and Temptation (Friðfinnsson), 293–94
probability, 13–14, 135–79
   of birthday matches, 137–40
   Fibonacci numbers and the stock market, 158–63
   and Friday the 13th, 135–37
   of matching two playing cards, 142–44
   mathematical logic of tic-tac-toe, 150–54
   mathematics of life insurance, 163–67
   Monty Hall problem, 150–54
   and selecting clothes, 141–42
   and statistics, 14, 163, 179
      use of central tendency, 172
   of various hands in poker games, 144–50

problem-solving strategies, 14
   being a detective investigating a case, 82
   mathematical applications in everyday life problems, 93–134
   problems with no solutions, 91
   and thinking "outside the box," 79–82
   use of extremes, 82–86
   working-backward strategy, 86–91
   See also logic
proportion
   in art, 12, 229, 289
   in the human body, 256, 284, 285
   mean proportion used to optimize soccer shots, 191–92
   See also proportional relationships
proportional relationships
   in interest, 166
   in life insurance, 165
   in the stock market, 162
Ptolemy (Claudius), 19
pyramids of Egypt, 12
Pythagorean means, 172
Pythagorean theorem, 19, 30, 196, 228, 278, 387
Pythagorean triples, 30

Quetelet, Lambert Adolphe Jacques, 255–56

Raimondi, Marcantonio, 295
rain
   getting wet in a faster or slower car, 83

INDEX 419

light rays striking raindrops, 262–71
rainbows, geometry of, 13, 258–73
   antisolar point of, 272
   creating an artificial rainbow, 273
   secondary rainbows, 259–60, 271, 272
   seeing a full-circle rainbow, 269, 272
Raphael, 289–91, 295, 300
ratios
   of basic units of length, area, and mass, 32
   Euler labeling ratio of diameter and circumference as π (pi), 144
   gear ratios in bicycles, 207, 208, 209, 211
   and geometric progression, 327
   and law of large numbers, 164
   and paper sizes, 106, 107, 108, 109
   phyllotaxis ratio, 252, 254
   in Reuleaux triangle, 122
   in Spirograph toy, 218–19, 221, 222, 223
   See also golden ratio (ϕ)
Rayleigh, Lord (John William Strutt), 342–44
reciprocals, 77, 88, 173, 277, 311, 333
rectangles
   maximum area of compared to minimum perimeter of, 112
   nets of a rectangular cuboid, 130–32
   people choosing a golden rect-angle as most pleasing, 279–81
   (*see also* golden rectangle)
reflections, use of
   in billiards, 201–206
   in installing an electric outlet, 116–17
   law of reflection (angle of reflection equal to angle of incidence), 262, 263, 346, 353
refraction, 227, 261, 264, 267
   angle of refraction, 261, 262
relativity, concepts of, 232–33
   general and special theories of relativity, 388–89
Rembrandt Harmenszoon van Rijn, 295
Reuleaux, Franz, 118, 119
Reuleaux pentagon, 123–24
Reuleaux triangle, 118–24
reverse reasoning. *See* working-backward strategy
Rhind, Alexander Henry, 21–22
Rhind Mathematical Papyrus, 22–23, 25
right-angle corner in ductwork, 111, 114
right-angled triangle used to compare three means geometrically, 177–79
right triangle
   comparing three means geometrically using a right-angled triangle, 177–79
   use of to compare altitude to hypotenuse, 174–75
Rodler, Hieronymus, 301
Roman calendars, 39

roman numerals, 15, 17, 60
  addition using, 37
  in check digits, 379
  history of, 34–39
  not having a place-value system, 30, 36, 38–39
  subtraction using, 37–38
  use of base-10 system, 30
Roosevelt, Franklin D., 136
Roosevelt, Theodore, 228
Rosenhouse, Jason, 158
roulettes (family of curves), 213
royal flush, probability of in poker games, 144–45, 150

St. Francis Preaching to the Birds (Giotto), 294
St. Jerome in His Study (Dürer), 302
St. Paul's Cathedral (London), whispering gallery effect, 342–44
Salt Lake City, UT, traffic lights, 362
Santa Maria delle Grazie (Milan, Italy), 297
satellite dishes, 350–51
satellite navigation and Global Positioning System, 381–89
Saunders, Richard, 40
SBN. *See* Standard Book Number (SBN system)
School of Athens (Raphael), 295, 300
Sebastián (Enrique Carbajal), 317
secondary rainbows, 259–60, 271, 272
Second Punic War, 349–50
seconds and timekeeping, 28, 29

secret messages, creating, 365–76
  keyword for translating a coded message, 371–76
secure data transfer, 376
Self-Portrait, A (Rembrandt Harmenszoon van Rijn), 295
semaphore traffic signals, 358
Sérusier, Paul, 295, 296
Seurat, Georges, 291–93
sexagesimal numbering system (base-60), 18
  and Babylonian mathematics, 18, 29–33
  and base-10 system, 18, 32
  handling fractions, 32
  measurements based on integer factors of 60, 32
  in timekeeping, 29, 33
  use of a symbol for zero, 19
shopping and mathematics
  combining coupons, 99–100
  comparing areas and perimeters, 110–13
  comparing prices and sizes, 93–94
  determining unit costs, 93–94
  evaluating coupons, 98–100
  paper sizes, 106–110
  rule of 72, 104–106
  successive percentages, 94–98
  understanding interest rates, 100–104
shot-putting, 187
şifr (empty space), 19
Signac, Paul, 295
simple interest, 100–101

INDEX   421

Sistine Madonna (Raphael), 289–90
Sixtus II (pope), 289–90
Sketchpad (Geometer software), 223
skewed tower, building, 328–34
Snellius, Willebrord, 261
Snell's law, 261–63
soccer, mathematical applications for optimizing a shot, 13, 187–92
socks, choosing, 141–42
solar reflector, 348–50
Sorensen, Jay, 125
sound waves and whispering galleries, 334–44
South Pole, 225, 230–32
space-time diagram for coordinated traffic lights, 362–64
speed
    and bicycles, 207–208, 210, 211
    calculating average speed, 168
    determining plane flights with and without winds, 169–70
    determining speed of a stream, 232–33
    determining speeds of cars, 85–86, 320, 359, 362, 363, 364
    and hands of a clock, 320
    of light, 261, 384, 385, 387, 388
    and trajectories, 182, 184, 185, 186–87
    wetness and speed, 82–83
spherical coordinates, 387
Spirograph toy, 212–23
sports, mathematical applications in, 13, 181–223
    best angle to throw a ball, 181–87

bicycles, 206–211
considering competitor's strengths and weaknesses, 81–82
mathematical strategies for billiards, 201–206
optimizing a soccer shot, 187–92
rowing and cork, determining stream's rate of speed, 232–33
single-elimination play and number of games needed, 80–82
tennis as game of angles, 192–201
understanding averages, 170–71
use of roman numerals, 34
square roots ($\sqrt{}$)
    square root of 2 ($\sqrt{2}$)
        and paper sizes, 108, 109
        and wrapping packages, 133
    taking the fifth root when averaging five items, 174–75
squares (geometric shape), 111, 112, 115, 186, 280
    not chosen for manhole covers, 118
    and wrapping packages, 131, 133
squares, magic. *See* magic squares
squaring numbers, 228
    with 5 as the terminal digit, 69–71
    in magic squares, 308
Standard Book Number (SBN system), 376
statistics and probability, 14, 163, 179
    use of central tendency, 172
statues, optimal position for viewing, 309–312

statute mile, 76
stock market and Fibonacci numbers, 158–63
  wave principle, 159–60
straight, probability of in poker games, 147, 150
straight flush, probability of in poker games, 145, 150
Straus, Ernst, 28
stream, determining rate of speed of, 232–33
Strutt, John William (aka Lord Rayleigh), 342–44
subtraction
  in ancient Egypt, 26
  in Roman mathematics, 37–38
  use of in creating divisibility rules, 56, 59, 60, 63–66
Super Bowls, use of roman numerals, 34, 35
synchronization of traffic signals, 356–65

Taft, William H., 140
tally marks, 15–17
Tannhäuser (Wagner), 137
technology and mathematical perspectives, 319–89
  building a skewed tower, 328–34
  coordination of traffic signals, 356–65
  creating caustic curves, 351–56
  creating secret codes and messages, 365–76
  Global Positioning System (GPS), 381–89

ISBN system, 376–81
paper folding, mathematics of, 322–28
parabolic concentrators, 348–49
parabolic reflectors
  flashlights, 345–48
  satellite dishes, 350–51
use of correction factor to determine movement of hands of a clock, 319–22
whispering galleries, 334–44
tennis as game of angles, 192–201
Terni Lapilli. *See* tic-tac-toe, mathematical logic of
tetrahedron, nets of, 130
Thorndike, Edward Lee, 281
three-digit numbers, 18, 71
three of a kind, probability of in poker games, 147–48, 150
throwing a ball, 181–87
thunderstorms, locating, 384–85
tic-tac-toe, mathematical logic of, 150–54
  variation that three Xs or Os in a row loses, 153–54
Tierney, John, 155
time dilation, 388–89
timekeeping
  attempts to decimalize time measures, 33
  history of, 28–29
  and sexagesimal numbering system, 29, 33
  See also clocks and watches

topology
- and the "Five-Bedroom-House Problem," 241–43
- and the Königsberg Bridges Problem, 237–41

Tour de France, 210, 211
tower, building a skewed, 328–34
traffic signals, coordination of, 356–65
trajectories, 184, 186, 187, 194–95, 215
- ballistic trajectory, 182–83

transcendental numbers, 144
transposition errors and ISBN numbers, 378, 379–81
triangles
- comparing three means geometrically using a right-angled triangle, 177–79
- Reuleaux triangle, 118–24
- right triangle, use of to compare altitude to hypotenuse, 174–75
- use of by Dürer, 285
- use of by Leonardo da Vinci, 288–89
- use of by Raphael, 289–90

Trinity (Masaccio), 294
triskaidekaphobia. *See* 13, as a lucky (or unlucky) number
Triumph of Galatea, The (Raphael), 295
Twelve Caesars, The (Suetonius), 367
two-digit numbers
- divisibility of, 62
  - by 4, 62
  - by 5, 62
  - by 8, 62
  - by 25, 62–63
- multiplication of, 73–74
  - by 9, 59
  - by 11, 57–58
- squaring of, 70

two pairs, probability of in poker games, 148, 150

Underweysung der Messung mit dem Zirckel und Richtscheyt (*Instructions for Measurement with Compass and Ruler*) (Dürer), 130, 301
unit costs, finding, 93–94
United Nations Headquarters, 12
United States Air Force, 383
unit fractions in ancient Egypt, 24–28
USTA National Tennis Center (renamed USTA Billie Jean King National Tennis Center), 193

Vigenère, Blaise de, and the Vigenère cipher, 369–70
Virgin of the Rocks, The (Leonardo da Vinci), 297–98
Vitruvian Man, 255, 284, 365–66
Vitruvius Pollio, 284
volume
- and paper folding, 325
- sizing a box for greatest volume, 115–16
- time needed to fill a tub with two hoses, 116

vos Savant, Marilyn, 150, 151

Wagner, Richard, 137
Wallis, John, 38

Washington, George, 52
watches. *See* clocks and watches
wave patterns and coordinated traffic lights, 356–65
Wave Principle, The (Elliott), 159
wave principle for the stock market, 159–60
Weber, Carl Maria von, 137
weighted sum, 378–79, 380
Weyden, Rogier van der, 294
whispering galleries, 334–44
"windy trips," determining time takes, 169–70
wine, concentrations of red and white, 84–85
working-backward strategy, 86–91
world. *See* Earth

"worst-case scenario." *See* extremes, use of to solve problems
wrenches, limitations for use of, 120–21

year, determining day for a date in any, 43–47

Zeising, Adolf, 256, 279
zephyrum, 19
zero (0), concept of, 17–20
    introduction into decimal system (base-10 system), 19
    lack of in Babylonian mathematics, 31
    lack of in Roman mathematics, 37